ACCOUNT OF THE

LIFE AND WRITINGS

OF

ROBERT SIMSON, M.D.

British Library Cataloguing-in-Publication Data
A catalogue record for this book is available from the
British Library

ACCOUNT OF THE

LIFE AND WRITINGS

OF

ROBERT SIMSON, M. D.

LATE PROFESSOR OF MATHEMATICS IN THE UNIVERSITY OF GLASGOW.

By the Rev. WILLIAM TRAIL, LL. D. F. R. S. Edin.

MEMBER OF THE ROYAL IRISH ACADEMY, AND CHANCELLOR OF ST. SAVIOUR'S, CONNOR.

Robertus Simson, M.D.

GEOMETRIAM SUB TYRANNO BARBARO SÆVA SERVITUTE DIU SQUALENTEM,
IN LIBERTATEM ET DECUS ANTIQUUM VINDICAVIT UNUS.

TO

THE REVEREND THE PRINCIPAL,

AND TO

THE PROFESSORS,

OF THE

COLLEGE AND UNIVERSITY OF GLASGOW,

THIS ACCOUNT OF THE LIFE

AND WRITINGS

OF A DISTINGUISHED MEMBER

OF THEIR SOCIETY

IS MOST RESPECTFULLY INSCRIBED.

b

CONTENTS.

ADVERTISEMENT.

A BOVE thirty years ago the late EARL STANHOPE honoured me with a requeſt to draw up an account of the Life and Writings of the late Dr. SIMSON, of Glaſgow, which might be publiſhed in the new edition of the *Biographia Britannica*. The ſlow progreſs of that great work left me much at liberty as to the time of preparing an article which could appear only near the end of it; and for a number of years having been occupied by engagements of a different kind, I was in ſome meaſure compelled to poſtpone the execution of my undertaking, much longer than I wiſhed to have done.

As there is not at preſent any near proſpect of the completion of the *Biographia*, I could not properly, at my time of life, defer any longer embracing the opportunity afforded me of paying this ſmall tribute of reſpect to the memory of that eminent man, by whoſe friendſhip and inſtruction I was honoured during ſome of the laſt years of his life. I thought it my duty therefore, though the undertaking was ſtill liable to interruptions from other concerns, to collect the neceſſary materials, and to arrange them for a ſeparate publication.

With reſpect to the incidents of Dr. SIMSON's life, particularly thoſe of the early part of it, I obtained ſatisfaɕtory information from a ſhort narrative drawn up by the Doɕtor's colleague and particular friend the late Mr. CLOW, which, from the intimacy ſubſiſting between them for many years, may be conſidered as authentic and accurate. From ſeveral gentlemen who had been in the ſame College with Doɕtor SIMSON other circumſtances have been communicated, and even from my own acquaintance with him, though only for a few years, ſome intereſting particulars came within the reach of my own obſervation.

With regard to Doɕtor SIMSON's ſcientific purſuits, there are abundant materials of information in his Works which have been publiſhed, and particularly in his learned Prefaces and Notes which accompany them. Some circumſtances, alſo, have been collected from his unpubliſhed MS. papers, and from the ſmall remains of his Correſpondence. It is much to be regretted that the greater part of his Mathematical Correſpondence, which appears to have been very extenſive during a great part of his long life, had been either loſt or deſtroyed in his own time. Some intereſting fragments of it, however, were found among his papers; and a ſeries of his letters to that eminent mathematician the late Earl STANHOPE, from 1750 to 1758, was in the moſt liberal and obliging manner communicated by the preſent Earl; and from them ſome curious notices of Dr. SIMSON's ſtudies have been extraɕted.

Some hiftory and explanation of the ancient geometrical Analyfis, almoft neceffarily, forms a part of an account of the literary life of Dr. Simson, by whom that Analyfis has been completely reftored and illuftrated. The *Mathematical Collections* of Pappus, in which is contained nearly every thing that is known from the Ancients of that celebrated inftrument of inveftigation, which they employed fo fuccefsfully in their geometrical inquiries, from the Doctor's early partiality for that branch of fcience, naturally became an object of his particular ftudy. Some account therefore of that work, and of Dr. Simson's commentaries on it, is requifite for giving a juft view of his geometrical labours, and for eftimating their importance to, what is fo much wanted, a corrected edition of that valuable Author. Even with the rifk of feveral repetitions, it has appeared convenient to detach from the Memoir the particular account of Pappus, and fome explanations of the ancient Analyfis connected with it, and to annex them as an Appendix. Thefe details, though uninterefting to many readers, may be acceptable to others; and my object in preparing them will be attained, if they can fave fome trouble to thofe who at any future time may undertake a new edition of the *Mathematical Collections.* Two fhort paffages of Pappus, concerning the ancient claffification of geometrical lines, are added in a fecond appendix, in the original language, with Commandine's verfion of them.

To thefe is fubjoined Dr. Simson's tranflation of the defcription of the Porifms of Euclid, in the Preface to the feventh book of Pappus; in which are fome material improvements of the former tranflations.

August, 1812.

Fig. 1

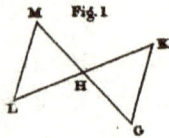

Fig. 2

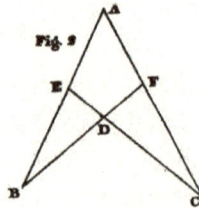

Fig. 3

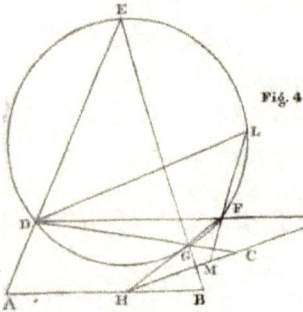

Fig. 4

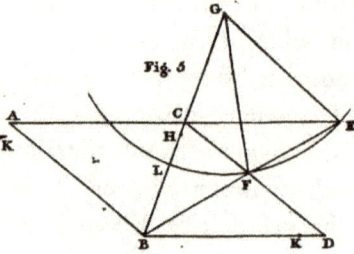

Fig. 5

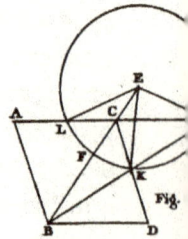

Fig. 6

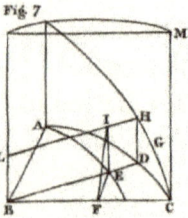

Fig 7

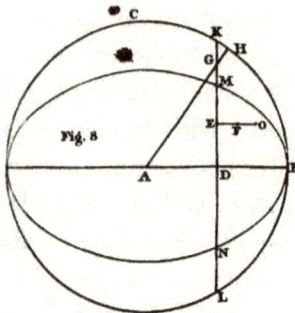

Fig. 8

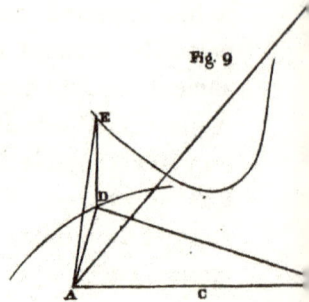

Fig. 9

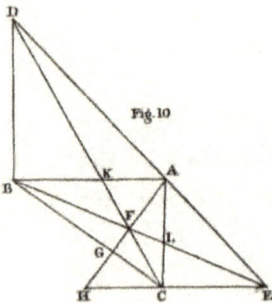

Fig. 10

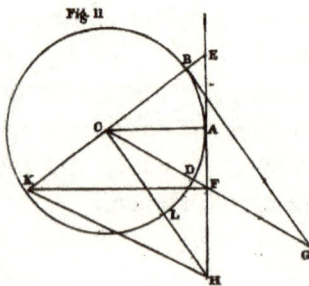

Fig. 11

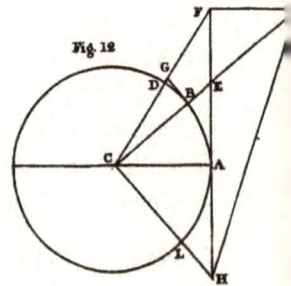

Fig. 12

SECTION I.

General Account of DR. SIMSON's *Life.*

DOCTOR ROBERT SIMSON, late Profeffor of Mathematics in the Univerfity of Glafgow, was the eldeft fon of John Simson, of Kirktonhill in Ayrfhire, and was born on the fourteenth of October 1687, O. S. Being defigned by his father for the Church, after having got the ufual fchool education, he was fent to the Univerfity of Glafgow, where he was diftinguifhed by his proficiency in claffical learning, and in the fciences.

At this time, from temporary circumftances, it happened, that no Mathematical Lectures were given in the College; but young SIMSON's inquifitive mind, from fome fortunate incident, having been directed to Geometry, he foon perceived the study of that fcience to be congenial to his tafte and capacity. This tafte however, from an apprehenfion that it might obftruct his application to fubjects more connected with the ftudy of theology, was anxioufly difcouraged by his father, though it would feem, with little effect.

Having procured a copy of EUCLID's Elements, with the aid only of a few preliminary explanations from fome more

advanced ftudents, he entered on the ftudy of that oldeft
and beft introduction to Mathematics. In a fhort time he
read and underftood the firft fix with the eleventh and
twelfth books ; and being delighted with the fimplicity of
language and accuracy of reafoning in Euclid, notwithftanding the difcouragements he met with, he perfevered in
his Mathematical purfuits;* and by his progrefs in the more
difficult branches he laid the foundation of his future eminence.
But though the bent of his inclination for thefe pursuits was
ftrong, he did not neglect the other fciences then taught in
the College ; and in proceeding through the regular courfe of
academical ftudy he acquired the principles of that
variety of knowledge, which he retained through life, and
which contributed much to the eftimation of his converfation and manners in fociety. His chief attention, however,
was directed to his favourite fcience; and his reputation as a
mathematician in a few years became fo high, and his general
character fo much refpected, that in 1710, when he was only
twenty-two years of age, the Members of the College, without
any folicitation on his part, made him an offer of the Mathematical Chair, in which a vacancy was in a fhort time expected
to take place. From his natural modefty, however, he felt
much reluctance, at fo early an age, to advance abruptly from
the fituation of a Student to that of a Profeffor in the fame
College ; and therefore he folicited permiffion to fpend one
year at leaft in London, where, befides other obvious advan-
tages, he might have opportunities of becoming acquainted

* In fome of his early MS. volumes ftill preferved, are notices of the books he had
been reading, and of which he had been making abftracts. In one of thefe, dated in
1705, (then in his eighteenth year,) are mentioned Oughtred's Clavis, Raphson's
Tracts, Jones's Synopfis, Kersey's Algebra, and feveral others.

with fome of the eminent mathematicians of England, who were then the moft diftinguifhed in Europe.   In this proper requeft he was readily indulged; and without delay he proceeded to London, where he remained about a year, diligently employed in the improvement of his mathematical knowledge. This journey turned out very favourable to his views; and he had much fatisfaction in the acquaintance of fome refpectable mathematicians, particularly of Mr. Jones, Mr. Caswell, and Mr. Ditton.   With the latter, indeed, who was then Mathematical Mafter of Chrift's-Hofpital, and well efteemed for his learning, he was more particularly connected; but from all of them he had opportunities of receiving information refpecting the progrefs of the fcience, both in England and on the continent of Europe.   When the vacancy in the Profefforfhip of Mathematics at Glafgow did occur in the following year, the Univerfity, while Mr. Simson was ftill in London, appointed him to fill it; and the Minute of Election, which is dated March 11th, 1711, concluded with this very proper condition: " That they will admit the faid Mr. Robert " Simson, providing always, that he give fatisfactory proof of " his fkill in mathematics previous to his admiffion."   He returned to Glafgow before the enfuing feffion of the College, and having gone through the form of a trial, by refolving a geometrical problem propofed to him, and alfo by giving " a " fatisfactory fpecimen of his fkill in Mathematics, and dex- " terity in teaching Geometry and Algebra;" having produced alfo refpectable certificates of his knowledge of the fcience from Mr. Caswell, and others, he was duly admitted Profeffor of Mathematics, on the 20th. of November of that year.

Mr. SIMSON immediately after his admiſſion entered on
the duties of his office, and his firſt occupation neceſſarily was
the arrangement of a proper courſe of inſtruction for the
ſtudents who attended his lectures, in two diſtinct claſſes. He
prepared elementary ſketches of ſome branches in which there
were not ſuitable treatiſes in general uſe. Among his papers
one of theſe ſtill remains, a tranſlation of the three firſt books of
L'HOSPITAL's Conic Sections, in which geometrical demonſtrations are ſubſtituted for the algebraical of the original,
according to Mr. SIMSON's early taſte on this ſubject. There
remain alſo ſome traces of his collections of Problems from
HUYGHENS, GREGORY, and others, in optics and aſtronomy,
for the uſe of his ſcholars; and it appears likewiſe that ſoon
after his admiſſion, he had given ſome public Lectures on the
Hiſtory of Mathematics, before the whole Univerſity, of
which ſeveral happen to be preſerved, and are proofs of the
extent and accuracy of his learning at the early period of life
when they were delivered.

Both from a ſenſe of duty and from inclination, he now
directed the whole of his attention to the ſtudy of mathematics;
and though he had a decided preference for geometry, which
continued through life, yet he did not devote himſelf to it,
to the excluſion of the other branches of mathematical ſcience;
and in the progreſs of this Memoir it will appear that he was
well acquainted with the modern analyſis, particularly as it ſtood
in the early part of his time. From 1711, he continued near
fifty years to teach mathematics to two ſeparate claſſes, at
different hours, five days in the week, during a continued
ſeſſion (or term) of ſeven months; beſides giving occaſional

inftruction which he was ever ready to communicate to thofe ftudents, who wifhed for more particular explanations of his lectures, or to make further progrefs in the ftudy of mathe-matics. Though the duties of a profeffor foon became familiar and eafy to him, yet they occupied a confiderable portion of his time, and divided it, fo as often to interrupt the courfe of his private ftudies.

His manner of teaching was uncommonly clear, and engaging to young people; and moft of his fcholars retained through life an affection and reverence for the Profeffor. The College of Glafgow in his time was in great repute both at home and abroad, to which Dr. HUTCHESON, Dr. MOOR, Mr. ADAM SMITH, and himfelf, much contributed. The refort of ftudents was great, and almoft all of them attended Dr. SIMSON's lectures. The knowledge of the elementary branches of mathematics, and of the moft ufeful applications of them, were thence much diffufed in the College, and fome tafte alfo for the ftudy of the higher branches was excited; but the early age of the greater number of the ftudents, their fhort refidence in College, and the neceffary appropriation of a confiderable portion of their time to other fciences, feldom admitted of that long and nearly exclufive cultivation of one particular fcience, by which alone, efpecially in mathematics, eminence ufually can be attained. Among Mr. SIMSON's fcholars, however, feveral rofe to diftinction as mathematicians. Dr. MATTHEW STEWART, who alone has applied the geometrical method of reafoning to the moft complex phyfical inveftigations, by univerfal acknowledgement, is to be named the firft. Mr.

WILLIAMSON,* a favourite pupil, from whom he had great expectations, died very young. Dr. JAMES MOOR, Greek Profeſſor at Glaſgow, and Profeſſor ROBISON,† of Edinburgh, were all well known as mathematicians of ſuperior abilities and attainments.

In the year 1758, Dr. SIMSON, being then ſeventy-one years of age, found it neceſſary to employ an aſſiſtant in teaching; and in 1761, on his recommendation, the Rev. Dr. WILLIAMSON was appointed his aſſiſtant and ſucceſſor.

The reſignation of Dr. SIMSON preſented an opportunity to the Principal and Profeſſors of recording in their minutes the affection which they felt for the Doctor, and their high admiration of his genius. A long paper for this purpoſe was drawn up by his colleagues, and it is expreſſed with all the warmth of attachment and reſpect, which it was natural for them to entertain for the father of the College, from whom the Univerſity had derived ſo much honour. Several of them had been his pupils, and all had lived with him in habits of friendſhip from the time of their becoming members of the Society. It is introduced in the following manner:

" The Univerſity Meeting do hereby gratefully, unanimouſly, " and warmly expreſs to Dr. SIMSON their moſt cordial thanks

---

* Mr. WILLIAMSON, afterwards the Rev. Dr. WILLIAMSON, Chaplain to the Britiſh Factory at Liſbon, in which ſtation he died.

† In the third and fourth editions of the Encyclopedia Britannica (in the article SIMSON R.) is a ſketch of Dr. SIMSON's life, which is known to be from the pen of Profeſſor ROBISON, of Edinburgh; and though, from ſome accidents not now to be aſcertained, there are ſome errors in dates, and in ſome ſmall circumſtances of the narrative, yet it bears the character of the diſtinguiſhed ability and knowledge of the writer.

" for his long, faithful, and eminent fervices to the Univerfity,
" in the courfe of fifty years; during which very uncommon
" period, he has with univerfal applaufe, been Profeffor of
" Mathematics here; to the great honour of this Univerfity,
" as well as to his own high reputation, which will laft as long
" as united elegance and fcience are admired among mankind."
It proceeds to mention generally fome of his moft diftinguifhed inventions, juftly ftating that he was the firft, in
modern times, who fully underftood and explained the ancient
analytical geometry; that he had already reftored, with fuperior accuracy and elegance, fome of the moft valuable works
of the ancient geometers, which for ages had been loft; and
that he purpofed to employ the remaining leifure of his life, in
compleating and publifhing others, not lefs neceffary for the
full illuftration of the nature and ufe of the ancient analyfis:
" for the accomplifhment of which valuable defign, (the
Minute concludes,) " of fo much importance to the advance-
" ment of true fcience, the Univerfity Meeting do wifh and
" pray that he may long enjoy the blefling of a vigorous old
" age; and they intreat Dr. SIMSON to be entirely affured,
" that they will at all times heartily embrace every opportu-
" nity of teftifying to him their gratitude, their affection,
" efteem, and veneration."
During the remaining ten years of his life, he enjoyed a
pretty equal ftate of good health; and continued to occupy
himfelf in correcting and arranging fome of his mathematical
papers, and occafionally for amufement, in the folution of
problems, and demonftration of theorems, which occurred from
his own ftudies, or from the fuggeftions of others. His con-

verfation on mathematical and other fubjects continued to be clear and accurate, yet he had fome ftrong impreffions of the decline of his memory, of which he frequently complained; and this probably protracted, and finally prevented his undertaking the publication of fome of his works, which were in fo advanced a ftate, that with little trouble they might have been compleated for the prefs.* His only publication, however, after the refignation of his office, was a new and improved edition of the Data of Euclid, which in 1762 was annexed to the fecond and corrected edition of the Elements. But from that period, though much folicited to bring forward fome of his other works on the ancient geometry, though he knew well how much it was defired, and though he was fully apprized of the univerfal curiofity excited refpecting his difcovery and illuftration of the Porifms of Euclid, he refifted every importunity on the fubject.†

* The following extracts from Dr. Simson's Letters to the Writer of this Memoir mention the impreffion he had of the decline of his memory:
     In a letter dated February 11th, 1766.——" As to the publifhing any thing on the " ancient geometry, give nobody any hopes of my doing it. You know my inability " to do any thing that requires fo much thought and application as that would do; " and if it be not by your affiftance in preparing them for the prefs, and in taking " care of the printing, I have fcarce any hopes of doing any thing that way, though I " much defire it." And in another letter of March 7th, 1767, he fays, " I wifh you " had not mentioned any thing that might give expectation of the Porifms from me, as " it is very doubtful whether I fhall be able to publifh any thing either about them, " or any other geometrical fubject; my memory, and confequently any other ability " for fuch things, being fo greatly decayed. However, your kind offer of affiftance " will make me give out fome figures to be cut, though I believe they cannot be " ready againft the time you propofe to be here."
     I fhall have afterwards occafion to mention the impreffion of a decline of memory which Dr. Simson feems to have felt many years before.
     † Sometimes he feems to have entertained ferious thoughts of publifhing the moft important of his remaining works, the Sectio Determinata, and the Porifms; and in

A life like Dr. SIMSON's, purely academical and perfectly uniform, rarely contains occurrences, the recording of which could be either interesting or useful. But his mathematical labours and inventions form the important part of his life, and supply the best illustration of his character; and with respect to them, there are abundant materials of information in his printed works; and some circumstances also may be gathered from a number of MS. papers which he left ; and which, by the direction of his executor, are deposited in the Library of the College of Glasgow.* It is to be regretted, that of the extensive correspondence which he carried on through life with many distinguished Mathematicians, a small portion only is preserved. The greater part of it had been lost or destroyed in his own time; so that in 1751, when he was desirous of reviewing some of his early letters to Dr. JURIN on a particular subject, he obtained a copy, probably from the archives of the Royal Society, of which Dr. JURIN at the time of the

one of his MS. books, of which the earliest date is of 1762, there is the following notice. " The title-page, if the book to which it belongs come to be printed, may " be thus:

" APOLLONII PERGÆI
" De Sectione Determinata Libri duo,
" Restituti."

" Quibus adjecta sunt non pauca Porismata; inter quæ habentur quædam EUCLIDIS, " quibus Doctrina Porismatum explicata et restituta est."

* By the liberal favour of the Principal and Professors, the Writer of this Memoir has had every accommodation for consulting these Papers, and also the very valuable Mathematical Library which Dr. SIMSON bequeathed to the University. He has received also from them some material communications respecting their distinguished Colleague; and he is particularly obliged to Professors JARDINE, YOUNG, and MILLAR, for procuring useful information, and aiding his researches.

C

correfpondence (1723) was fecretary. Through Dr. JURIN he had fome intercourfe with Dr. HALLEY; and both at that time, and afterwards, he had frequent correfpondence with Mr. MACLAURIN, with Mr. JAMES STIRLING, Dr. JAMES MOOR, Mr. WILLIAMSON of Lifbon, and more particularly with the Rev. Dr. MATTHEW STEWART; of which there are fome notices in his printed works, and of which alfo there are fome remains among his unpublifhed papers. In the latter part of his life, his mathematical correfpondence was chiefly with that eminent Geometer the late Earl STANHOPE,* and with GEORGE LEWIS SCOTT, efq; then a Commiffioner of Excife, and well known for his fcientific attainments; and from Dr. SIMSON's letters to both thefe gentlemen, fome illuftration of his opinions and purfuits will be obtained, which fhall be taken notice of in a fubfequent part of this Memoir.

* The Doctor's affectionate refpect for that venerable Nobleman is ftrongly marked by the concern which he expreffes in a letter to himfelf on hearing a report of his being indifpofed. "May it pleafe the gracious GOD to preferve fo valuable a " life for many good purpofes, both with refpect to your Lordfhip's honourable family, " and the public good; and particularly for promoting of real and ufeful knowledge " of every kind."——Letter dated 7th March 1755.

SECTION II.

Particular Account of Dr. SIMSON's *Mathematical Studies, and of the Works publiſhed by himſelf.*

IT has already been obſerved, that Dr. SIMSON directed his early attention to the ancient Geometry. Dr. HALLEY, by his edition of the Conics of APOLLONIUS, and more particularly by his publication of two Treatiſes of APOLLONIUS, accompanied with a corrected edition of the preface to the ſeventh Book of PAPPUS, gave conſiderable aid to thoſe who embarked in that ſtudy. The *Mathematical Collections* of PAPPUS are indeed the chief repoſitory of information reſpecting the Geometry of the Ancients, and eſpecially reſpecting their analyſis, which has been the ſubject of much diſcuſſion among the moderns. This moſt intereſting work contains ſome curious mathematical hiſtory of former times, but is more particularly valuable, by the account, contained in the preface to the ſeventh Book, of the treatiſes of the analytical geometry of the ancients, which together obtained the name of τόπος ἀναλυομένος. In this preface there is, firſt, a general expoſition of the analyſis employed by the ancients, both in the ſolution of problems, and in the demonſtration of theorems; their follows a particular deſcription of the nature and contents of a certain number of theſe treatiſes, which we may preſume were confidered by PAPPUS as the moft important; and an enumeration of the whole is added, confifting of thirty-three books. The feventh book of PAPPUS itfelf confifts of a number of Lemmata, or fubfidiaty propofitions, not contained in EUCLID, but affumed or employed in the feveral treatifes which are fo fully defcribed in the preface.

It is through the medium of COMMANDINE's tranflation only, that the importance of this work of PAPPUS is yet known to the public. Though many manufcript copies of the original remain in various libraries of Europe, it has never been printed; and no attempt has been made to correct the tranflation of COMMANDINE, whofe zeal, learning, and ability, in promoting the knowledge of the ancient Geometricians by ufeful tranflations, are very meritorious; though this pofthumous verfion of PAPPUS, no doubt from its not having received his final corrections, remains in a lefs perfect ftate.*

Dr. SIMSON, however, ferioufly applied himfelf to the ftudy of this tranflation; and notwithftanding its unfinifhed ftate, it was the means of enabling him to inveftigate, and fully to illuftrate the principles, of the analytical geometry of the ancients. The generality of modern mathematicians, from not having fully confidered the intimations of PAPPUS on this fubject, fell into ftrange mifapprebenfions refpecting it.†
When they examined the works of the ancient geometers,

* For a particular account of PAPPUS, and of this tranflation, fee the Appendix.

† VIETA muft be excepted; who, though he did not ufe the ancient analyfis in his *Apollonius Gallus*, yet he feems perfuaded, that the ancients had a true geometrical analyfis. For in his firft Appendix to that Treatife (p. 339, Op. VIETÆ) he obferves, "At Algebra quam tradidere, THEON, APOLLONIUS, PAPPUS, et alii véteres analyftæ omnino geometrica eft;" &c.

Euclid, Archimedes, and Apollonius, they admired the
elegance and clearnefs of their demonftrations, but wondered
by what contrivance they had invented and proved fo many
curious theorems, and obtained folutions of fo many difficult
problems. They afferted even that the ancients muft have
poffcffed an analyfis equivalent to the algebraical; but that
they had induftrioufly concealed it, in order to excite the
greater admiration of their inventions. This opinion, naturally
improbable, is however unrefervedly expreffed by fome eminent
mathematicians; by Francis Schooten, by his brother Peter
Schooten, by Peter Nonius or Nunez; and what is more
remarkable, is avowed in later times by the profound Dr.
Wallis,* who was certainly acquainted with Pappus, both
in the original, and in Commandine's tranflation. But the
important communications of that author refpecting the geometrical analyfis of the ancients feem not to have been duly
confidered by this very eminent Mathematician, though he
publifhed, with learned and valuable notes, a fragment of
Pappus, not contained in Commandine's tranflation, but
which he difcovered in one of the two Savilian MSS. of the
Mathematical Collections.

Dr. Barrow alfo, in his Lectures,§ gives countenance to
this erroneous judgment on the ancients, which has been

* Wallis's Algebra, chap. ii.

§ In Dr. Barrow's four Lectures on the Difcoveries of Archimedes, after ftating
his plan in p. 341, he adds, "unde patebit qualem analyfin, et quam noftræ modernæ
"fimilem exercuerit." And in p. 376, after giving an algebraical analyfis of a
problem, (viz. Prop. v. Lib. 2. Archimed. de Sphæra et Cylindro,) producing a
cubic equation, he derives from it the proportion given by Archimedes, in which a
certain ftraight line is to be cut in order to refolve the problem; he adds, "Qui ipfiffi-

pronounced by fo many mathematicians of modern times; and even fo late as the publication of Sir Isaac Newton's Fluxions in 1736, Mr. Colson, the commentator on that work, ftates this to be the opinion of "many of our modern geo- "metricians."‡

This charge, however, againft the ancients, as is juftly obferved by Dr. Simson,§ was altogether groundlefs. The only algebra known to the ancients was that of Diophantus,

"mus eft analogifmus ifte, ad quem rem deduxit Archimedes; quod ipfum fatis "prodit et arguit, qualem is analyfin ufurparit. Nam huc eum deveniffe, varias iftas "proportionum compofitiones, divifiones, permutationes, ac inverfiones, quales in dif- "curfu fuo oftentat adhibendo, pene fupra fidem eft." But in this cafe the ancient analyfis is properly applicable; and under the management of Archimedes, its power in this propofition need not excite furprize. He refers alfo to the elegant fynthetic folution of this problem by Hutonens, by means of the trifection of an arch. Huyg. Illuftr. Probl. prob. 1. Perhaps fome of the difcoveries of Archimedes may have been fuggefted to him by views and reafonings refembling thofe of Cavalerius, as is conjectured by his Commentator Torelli; but he eftablifhed their truth by rigorous demonftrations, which might be naturally derived from the only analyfis of ancient times, the geometrical.

‡ See Colson's Commentary on Newton's Fluxions, p. 143: and it may be here remarked, that the obfervations of Sir Isaac Newton, (page 1,) to which this part of the Commentary refers, feem, to imply, that he confidered the ancient geometry, which he greatly admired, as fynthetical only, without having much confidered the nature and merits of their analyfis. Some of his obfervations in the Arithmetica Univerfalis naturally lead to the fame inference. See fect. iv. cap. 1. art. 18: alfo, Appendix de Æquationum Conftructione Lineari, art. 51. Dr. Pemberton's account of Sir Isaac Newton's opinion of the ancient geometry, of his regret that he had not ftudied it more particularly, and alfo of his remarks on De Omerique, corresponds with thefe obfervations in Sir Isaac's works. It may be added, that opinions like that of Mr. Colson are to be found in many refpectable recent writers; and confiftently with them, the ancient geometry is generally diftinguifhed as synthetical, while the term analysis is almoft exclufively applied to the modern fyftem,

§ See the Preface to Dr. Simson's reftoration of the Loci Plani of Apollonius, where this opinion is particularly ftated, with the Doctor's fatisfactory refutation of it.

which was never applied to geometry. But they had an
analyfis of their own, of which they made no fecret, and of
which there are fome fhort fpecimens, even in EUCLID's
Elements.* This analyfis however is more fully defcribed by
PAPPUS, who alfo obferves, that in order to facilitate the
folution of geometrical problems by this method, the ancients
compofed no lefs than thirty-three books, which collected
together got the title of τόπος ἀναλυομίνος; and of the chief of
which, as has already been mentioned, he gives an interefting
account in the preface to his feventh book.

What led probably to this prevailing miftake of modern
mathematicians was, that in the moft valuable ancient treatifes
ftill remaining, fuch as thofe of ARCHIMEDES, and even the
Conics of APOLLONIUS, the analyfis of their propofitions
(particularly of their theorems) is generally omitted ; though
there can be no doubt that this method was employed by
them, both in refolving problems, and for afcertaining the
truth of theorems, which now appear in their works only in
the fynthetic form, and which had been either propofed to
them by others, or had occurred to them in their own ftudies.
In copying their books, however, for general ufe, for the fake
of fhortnefs, (which, before the invention of printing, was an
object of confequence,) and even for facility to the reader
in acquiring the knowledge of their difcoveries, the analyfis
was often omitted. In the fifth book of PAPPUS is an example
of this omiffion, for which he affigns the reafon juft now
mentioned, and which we may believe influenced the
other more ancient geometers. He obferves, that in the com-

* EUCLID's Elem. book xiii. prop. 1, 2, 3, 4, 5.

parifon which he is about to give of the five regular folids·
having equal furfaces, for the fake of *brevity* and *perfpicuity*,
he is to employ only the fynthetic method, and not the
analytic, which fome of the ancients had ufed in treating that
fubject.† But in the other analytical treatifes of APOLLONIUS,
containing folutions of general problems for facilitating the
refolution of any particular geometrical problem which can
be reduced to a cafe of them, a full analyfis of every cafe is
given, being effential to this important application of fuch
problems. In the *Sectio Rationis,* recovered from the Arabic
by Dr. HALLEY, is a fpecimen of this compleat manner of
folution, which, we muft prefume, was followed in the other
treatifes, now unfortunately loft.

Thefe books appear to have been all exifting in the time of
PAPPUS, but the greater part of them have fince perifhed, or
at leaft they are not known to remain, either in the original,
or in any tranflation. The book of *Data* by EUCLID, in the
Greek ; two books of *Sectio Rationis,* by APOLLONIUS, in an
Arabic verfion ; and feven books of the *Conics* of APOLLONIUS,
four in Greek and three in Arabic ; are all that have been preferved. But fortunately the defcriptions by PAPPUS of eleven
more of thefe books are fo particular and entire, that fome eminent modern mathematicians have been able to reftore them,
with various fuccefs indeed, as fhall be afterwards more particularly ftated. Thofe by Dr. HALLEY, from his fuperior tafte and
knowledge of the fubject, are in the pureft ftyle of geometry.

† His expreffion is, " Καὶ τὴν ἔφοδον τῶν ἀποδείξεων ἰχύσας, ἢ διὰ τῆς ἀναλυτικῆς λεγομένης
" θεωρίας δὲ ἧς ἔνιοι τῶν παλαίων ἐποιοῦντο τὰς ἀποδείξεις τῶν προειρημένων σχημάτων, ἀλλὰ διὰ
" τῆς καλὰ σύνθεσιν ἀγωγῆς ἐπὶ τὸ σαφίστερον καὶ συληπτότερον ὑπ' ἐμοῦ, διασκευάσμισας." &c.
Ex Cod. BULLIALDI, ad fol. 99. a. Com.

The others, by VIETA, SNELLIUS, MARINUS GHETALDUS, FER-
MAT, SCHOOTEN, and WALLIS, of which fome are algebraical,
and fome geometrical, though refpectable works of ingenious
men, are defective in that elegance and compleatnefs of
folution which diftinguifh the analytical writings of the ancient
Geometers. A field was thus ftill open, for a perfon of Dr.
SIMSON's tafte and genius, for attempting a more perfect reftoration of thefe curious works of antiquity; and it appears, that
foon after he was placed in the Mathematical Chair, he fet
about the inveftigation. From his papers ftill remaining, we
learn that his endeavours were firft directed to the improvement of the defective reftorations of these books, by preceding
geometers. From the fame fource alfo we know, that within
a few years he turned his attention to the Porifms of EUCLID;
and it was to be expected that the curious nature of thefe
Propofitions, and even the difficulty of the inveftigation, would
attract the early notice of his ardent mind.

Unfortunately the defcription of the Porifms by PAPPUS,
in all the MSS. of his *Collections* which have been examined,
is fo mutilated, that every attempt to reftore them, before
Dr. SIMSON's time, had failed. The firft part of the defcription, which feems to be entire, is calculated only to
excite curiofity; being too general for conveying any precife
notion of thefe Propofitions, or for giving any effectual affiftance for the recovery of them: and the remainder, containing
a detail of the contents of EUCLID's work, is through the
whole fo depraved by the injuries of time, that all endeavours to explain it were nugatory. Some Geometers
indeed, of great name, flattered themfelves that they had got

D

poffeffion of the fecret of this peculiar clafs of Propofitions;
but fubfequent to them, Dr. HALLEY, with all his genius, his
extenfive knowledge, and his fuccefsful experience in unra-
velling fome other pieces of ancient geometry, gives up the
Porifms as a hopelefs purfuit; and he admits that the defcription of PAPPUS, as it now ftands, is unintelligible and ufelefs.*
Dr. SIMSON, however, fully proved that thofe ingenious
men before Dr. HALLEY, who fuppofed that they had difcovered the Porifms, had certainly deceived themfelves; and
that the celebrated FERMAT alone, in modern times, had
acquired fome notion of the nature of Porifms, but without
being able to unfold it compleatly; and without having reftored
any one Propofition, that could be fuppofed to belong to the
Treatife of EUCLID.

The Doctor himfelf candidly informs us, how long and how
ferioufly he had ineffectually laboured in fearch of this ænigma,
as the nature of a Porifm truly became, from the very mutilated ftate of the only exifting defcription of it. An ardent
curiofity was a prominent and well-known part of his character;
and in this cafe, his curiofity was enlivened by his predilection
for the ftudy of the ancient geometry; but we may prefume alfo,
that a natural and laudable ambition for the diftinction which
would refult from fuccefs, in a purfuit in which the greateft
Mathematicians of his own and of the preceding age had failed,
would animate his zeal and perfeverance in the invefligation.
He had been occupied on this fubject even in the year 1715;

* " Hactenus Porifmatum Defcriptio, nec mihi intellecta nec lectori profutura.".
See the Preface to the Seventh Book of PAPPUS, edited by Dr. HALLEY, octavo, 1706,
page 87, note at the end of the Porifms.

and perhaps before that period; for he obferves, that in that
year he had demonftrated the firft cafe of FERMAT's fourth
Porifm, before he had acquired the knowledge of the nature
of that clafs of Propofitions.*

The firft direction of his refearch feems to have been, to endeavour to difcover the Porifms, from the general defcription of
them in the beginning of the account given by PAPPUS; and
when this failed, he tried to recover fome of the individual
Porifms, from which he hoped to afcertain the diftinctive
qualities of thefe Propofitions; but in this attempt he had no
better fuccefs. He continued his refearches, and devoted his
whole attention to the fubject. For a confiderable time his
imagination was compleatly occupied by it: his mind was
harraffed by the conftant, but unfuccefsful exertion; he loft
his fleep, and his heath was injured: but all his endeavours
were ineffectual; and therefore he finally determined to banifh
for ever the fubject from his thoughts.

For fome time he maintained this refolution, and applied
himfelf to other purfuits;† but afterwards he happened to be
walking with fome friends on the banks of the river Clyde at

* SIMSON's Pofthumous Works, p. 540, laft line.

† It is not improbable, that Dr. SIMSON about this time turned his attention to
Algebra, and that, on defpairing of the Porifms, he had employed himfelf in the
inveftigation of Seriefes for the Circle, of which an account is given in this Memoir,
(fee Note H.) The Porifms, as I reckon, were firft difcovered in April 1722. For
fome time after, it is reafonable to fuppofe that he was almoft entirely occupied in
profecuting his invention; and therefore, as he tranfmitted thefe Seriefes to Dr. JURIN
in February 1723, and from his letter alfo it appears that they had been found out
fome time before, it is very probable they were inveftigated before his invention of the
Porifms. They may be confidered as a refpectable fpecimen of what Dr. SIMSON might
have done in Algebra, had he devoted his attention to that branch of Mathematics.

Glafgow, and by accident being left behind his company, he inadvertently fell into a reverie refpecting the Porifms. Some new ideas ftruck his mind, and with his chalk having drawn fome lines on an adjoining tree, at that moment, for the firft time, he acquired a juft notion of one of EUCLID's Porifms.* I have repeatedly heard him relate the occurrence, which he feemed to do with pleafure, and he mentioned even the fcite of the tree on which he defcribed the fortunate diagram.†

After this firft difcovery, however, it required time and much inveftigation, before he could reftore, to his fatisfaction, the general Propofition of EUCLID's firft book of Porifms; and it appears that his firft communication of the difcovery to the Mathematicians of London was through Mr. MACLAURIN. Juft before Mr. MACLAURIN's fetting out from Scotland for France, by the way of London, in 1723, Dr. SIMSON communicated to him the Propofition he had recently reftored, and which a fhort time after was printed in the Philofophical Tranfactions. It appears alfo that Dr. JURIN, then Secretary

* This incident is generally ftated by Dr. SIMSON himfelf, in his Preface to the Porifms. Poft. Works, pp. 319, 320.

† Though moft of Dr. SIMSON's notes and propofitions are dated, yet there is no entry to be found among his papers of this particular incident. From the dates of feveral of his firft inveftigations of the Porifms, both in his PAPPUS, and in other MSS. it had, with much probability, occurred in April 1722; for in that month are fome of his firft notices about the Porifms. At the end of one of them he adds, " Hodie hæc " de Porifmatis inveni, R. S. April 25, 1722." Alfo in a note on Prop. 131, lib. vii. PAPPI, he fays, " Poftquam vero ipfa Porifmata, ea velim quæ generali propofitione " in præfatione ad hunc librum feptimum, complexus eft PAPPUS, multo labore at fuc- " ceffu præter quem fperare æquum fuit felici, tandem ex manca et contracta " admodum PAPPI defcriptione inveftigavimus; facilius erit lemmata hifce infer- " vientia, ad priftinum nitorem reftituere. April 27, 1722. ROB. SIMSON."

of the Royal Society, had mentioned in a letter to Dr. Simson
this communication from Mr. Maclaurin; for Dr. Simson,
in a letter to Dr. Jurin, dated Feb. 1, 1723, in reply to one
he had just received from him, fays, "The Propofition of
" Pappus, which I fhewed Mr. Maclaurin when here,* is
" the general Propofition into which Pappus collected the
" Porifms of the firft book of Euclid's Three; which, together
" with the more general Propofition immediately fubjoined,
" and which are both deficient, (I mean imperfect,) I have,
" with no little inveftigation, recovered and demonftrated,
" together with fome few of the Porifms; the reft I have not had
" leifure enough to try. I mean thofe of the firft book, for as
" to thofe of the two others, excepting what may be included
" in the fecond of the above-mentioned Propofitions, I believe
" it will be extremely difficult for any body to reftore them. If
" I underftand you like to fee them, I fhall fend them as foon
" as I can get the fcribbles I have about them wrote over fair."†
It would appear that the refult of this letter was an immediate communication of the Propofitions to Dr. Jurin, which,
as might be expected, were much approved by thofe eminent
Geometers Dr. Halley, Mr. Machin, and Dr. Jurin himfelf. In the courfe of the fame year the Paper was printed
in the Philofophical Tranfactions,‡ and Dr. Simson, in a letter

* This is mentioned in the Preface to Dr. Simson's Conics, p. vi. 2d edit. See
alfo Note A. at the end.

† The principal object of this letter was to communicate to Dr. Jurin fome Seriefes
about the Circle, of which notice will be taken afterwards. See alfo Note H. at
the end.

‡ Vol. XL. for 1723, p. 330. In the fame volume of the Tranfactions, p. 248, is
a Paper by Dr. Pemberton, containing fome propofitions about the Rainbow; at the

to Dr. JURIN, of January 10, 1724, expreffes his fatisfaction at
the reception of his communication in the following manner:
" The honour you have done me in printing the Paper I fent
" up, in the Philofophical Tranfactions, is what I am very
" fenfible of ; and you may be fure the approbation any thing
" in it has had from fuch good judges as Dr. HALLEY, Mr.
" MACHIN, yourfelf, and the other learned gentlemen you
" were pleafed to fhew it to, cannot but give me a great deal
" of pleafure; and nothing could excite me more to endeavour
" to reftore the other Porifms, of which there are the leaft
" data: but I find it a very difficult affair, efpecially to
" one who is fo flow, as by much experience I find myfelf.
" However I fhall at leifure hours, GOD willing, try what I
" can do. I defire you may give my humble refpects to Dr.
" HALLEY and Mr. MACHIN; and be pleafed to tell the Doctor,
" that as to the meaning of the words E'àν ὕστιϛ ἢ παρυπ]ίϛ,
" ἢ παραλλήλϛ, I never was folicitous about the meaning of
" the propofition, but at his defire I have confidered the paf-
" fage," &c.   Dr. SIMSON then enters into a long and learned
difcuffion of the proper meaning of this paffage, of which the
refult is fhortly ftated in a note, page 348, Poft. Works.*

beginning of which he obferves, " For the greater brevity, I fhall deliver them under
" the form of Porifms; as, in my opinion, the ancients called all Propofitions treated
" by analyfis only." This intimation naturally gave fome diffatifaction to Dr. SIMSON,
as affuming the appearance of being the firft to announce to the world the nature of the
ancient Porifms.   In the copy of this volume of the Tranfactions belonging to the
College Library of Glafgow, the Doctor wrote a fhort animadverfion on the margin
of the page, which concludes with thefe words: " but he (Dr. PEMBERTON) has
" entirely miftaken the nature of a Porifm, and his two Propofitions have neither the
" form nor matter of Porifms. R. S."

   * The reft of the letter may be interefting to fome readers, and is therefore placed
in Note C, at the end.

For a few years after his difcovery of the Porifms, his mind
was much engaged in the further profecution and illuftration of
it. It appears, however, from his papers, that at this very time
he occafionally applied to other branches both of ancient and
modern Mathematics; though, as was to be expected, the chief
object of his ftudy was the improvement and extenfion of his
inveftigation of the Porifms, by recovering more of EUCLID's
Propofitions, and adding alfo fome of his own, and fubfequently
alfo the contributions of a few mathematical friends, to whom
he communicated the interefting intelligence of his difcovery.
There are many indications of his intention of publifhing the
Porifms; but from various caufes he poftponed the execution
of it, till, in the progrefs of life, he had acquired fo very ftrong an
impreffion of the decline of his faculties, that he reluctantly gave
up the defign. The *Treatife on Porifms*, however, with fome
other valuable tracts, were publifhed a few years after the
Doctor's death, at the fole expenfe of his highly refpected and
learned friend the late Earl STANHOPE; and by the full explanation of the Porifms contained in that volume, (of which I fhall
have occafion to give a more particular account,) this hitherto
inexplicable portion of ancient geometry was, by the Doctor's
perfeverance and ingenuity, compleatly reftored and explained.
It may juftly be admitted, that the fatisfaction and the pride
of invention, in the elucidation of this very difficult branch of
ancient fcience, involved in fuch obfcurity from the depraved
ftate of the only remaining account of it, were properly high;
and by liberal and fcientific minds, the warmth with which he
expreffes thefe feelings will not be difapproved: " Defcriptio
" autem quam tradit (PAPPUS) Porifmatum adeo brevis eft et

" obfcura, et injuria temporis aut aliter vitiata, ut nifi Deus
" benigne animum et vires dederat in ea pertinaciter inquirere,
" in perpetuum forfan geometras latuiffent."*

It has been already obferved, that before Dr. Simson obtained
any juft notion of the Porifms, he had begun to improve the
reftorations of other ancient geometrical treatifes; which had
been attempted by preceding geometers, particularly of the
Loci Plani, and *Sectio Determinata*,† of Apollonius. To
thofe who have fome knowledge of this fubjeft, it is needlefs
to explain the value of thefe two treatifes; particularly of
the former, as containing many elegant Theorems, and as being
eminently ufeful in the refolution of Problems. The *Loci
Plani* of Apollonius had been reftored in a certain manner,
both by Fermat, and by Francis Schooten. In the pofthumous works of the former very diftinguifhed Mathematician,‡
is given a reftoration, geometrical indeed, but fynthetical only
without analyfis, and deficient alfo in other material points,
particularly in the diftinction of the cafes, and in afcertaining
the determinations; without which no geometrical refolution
can be confidered as compleat.§

* Opera Reliqua, p. 513.

† The *Sectio Determinata* was not publifhed in his own time, and I therefore defer
any notice of it, till an account be given of the volume of Pofthumous Works.

‡ Fermat had reftored this work of Apollonius before 1629, as appears by a
letter of his, p. 153 of his Works. But it was printed only in 1679, among his
Pofthumous Works. See Fermat. Oper. Varia, tom. 11. p. 12. Fermat had
not feen the Preface to the 7th book of Pappus in Greek.

§ The ufe of *Loci* in the refolution of Problems is very obvious, even to thofe who
are but little acquainted with ancient Geometry. The great importance of the
compleat diftinction of cafes and determinations in thofe treatifes of the τόπος ἀναλυ-

The reftoration by Schooten has fimilar defeds; in a few only of the problems an analyfis is given, but one purely algebraical; and he acknowledges in his preface* that his reftoration of the *Loci Plani* was defigned to be an illuftration of the geometry of Des Cartes, by furnifhing proper examples of his method. Though it appears that Dr. Simson had begun, at a very early period, to reftore the *Loci Plani* after the ancient model, and though the work was almoft compleated before he publifhed his *Conic Sections* in 1735, yet it was not printed till 1749. What reafons occurred for this delay cannot now be conclufively afcertained; but we learn from his papers, and from fome remains of his correfpondence, that at one time he had defigned to add one or two books of *Loci* to thofe of Apollonius. There are many detached Propofitions, and even fome Seriefes of Propofitions of that defcription, in which the two books of Apollonius are quoted as part of the fame work, which fufficiently afcertains his purpofe, at the time of writing them.† It is indeed much to be regretted, that he did not purfue this

ομίνος, which confift of general Problems, to which other Problems may often be reduced, will be explained afterwards. The fame attention to the cafes and deter-mination of *Loci*, and alfo of Porifms, is equally neceffary, to render the application of them ufeful to the folution of Problems.

* F. Schootenii *Exercitationes Mathematicæ*, Præf. 1657.

† In a letter to Earl Stanhope, dated September 10, 1750, after mentioning that the *Loci Plani* were juft printed, and about to be publifhed, he adds, " that he once " defigned to have added feveral other *Loci Plani*, but thought it beft now to give " only thofe mentioned by Pappus, with a few to make fome of them more com-" pleat."——There is alfo among his papers a fketch of a title for the *Loci Plani*, comprehending fome additions to thofe of Apollonius.

E

scheme; as no work could be more generally useful in geometrical investigations, than an extensive and well-arranged collection of *Loci*. The Doctor's plan seems also to have extended to solid *Loci*, of which indeed he gives a hint, in the preface to his *Conic Sections*; and among his papers are also some small Sets of such Propofitions, which indicate the probability of the defign which I fuppofe him to have entertained.* The hefitation about making additions to the Treatife of APOLLO-NIUS, probably contributed to the delay in printing the work, which was not executed till 1749. He then met with fome unexpected difficulties in treating with a bookfeller for the fale of the whole impreffion, which alone prevented the publication at that time;† and except a few copies diftributed among his friends in 1750, the book remained unpublifhed, till after his death. Such is the elegance of method, and the ingenious contrivance of demonftration in this work, that he has truly exhibited a copy, or at leaft fo very nearly a copy, of the work of APOLLONIUS, that little regret need be had for the lofs of the original. The preface alfo is well deferving the attention of thofe who wifh to acquire juft notions of the ancient books of analyfis.

The firft publication by Dr. SIMSON, except the Paper on Porifms in the Philofophical Tranfactions, was his Treatife *Sectionum Conicarum, libri v.* which appeared in 1735.

* In a great many of the Propofitions of folid *Loci* the correfponding algebraical equations are added. From fome obfervations in the preface to the feventh book of PAPPUS, it appears that the ancients had confidered feveral varieties of *Loci*, particularly the plane, the folid, the linear, and thofe called *Loci ad medietates*, arifing from mean proportionals; but of thefe laft only a very few imperfect notices remain in the beginning of the third book.

† This account is given by Dr. SIMSON, in a letter to the late Earl STANHOPE, Sept. 10th, 1750.

He had obferved, in the firft years of his ftudy of Mathematics, that the Treatifes on Conic Sections, then in moft
general ufe and eftimation, were entirely algebraical; and the
great merit of the work, written in that ftile by the Marquis
DE L'HOSPITAL, contributed not a little to the popularity
of this mode of treating geometrical fubjects. It occurred
therefore to Dr. SIMSON, that a treatife on Conic Sections,
written on the purer model of antiquity, might have fome
influence in correcting the prevailing falfe tafte, of introducing
algebraical calculation into thofe branches of geometry
where it was not neceffary, and where it fupplanted a more
elegant form of analyfis and demonftration. To exhibit,
therefore, a juft comparifon of the two methods, he affumed
the fame definitions of the Conic Sections, as L'HOSPITAL and
others before him had employed; and from them, with the
true fimplicity and accuracy of the ancient fchool, he
deduced not only the properties of thefe curves as given by
all preceding writers, but added many new and important
Propofitions of his own,* with the generalization and improvement of many, which had been previoufly difcovered.
It is unneceffary to enter into any particular difcuffion of

---

* The precife object which Dr. SIMSON had in view, was the occafion of
his adopting the definitions of the curves given by L'HOSPITAL and others,
in which the language is rather mechanical, than of a ftrictly fcientific character.
This I have frequently heard the Doctor exprefs; and he obferved at the fame time,
that he confidered the derivation of the properties of thefe curves from the cone, after
the original method of the ancients, as the beft; and that if a definition of them,
from a defcription in a plane, were to be affumed as moft expedient, a more correct
form of language than L'HOSPITAL's ought to be ufed; which indeed is generally
done by thofe modern geometers who have relinquifhed the confideration of the cone,
in defining this clafs of curves.

the merits of a work, which has been for fo many years before
the public, which was received with general approbation
on its appearance, and which, notwithftanding more recent
improvements, ftill maintains the reputation, of being the beft
example of the ancient method of demonftration, of having
communicated large additions to the theory of the Conic
Sections as it ftood in his time, and of exhibiting a clear and
juft arrangement of the properties of thefe curves.

This Treatife became a part of Dr. SIMSON's Courfe of
Lectures in the College; it was reprinted in 1750 with feveral
additions, including fome valuable communications by Dr.
MATTHEW STEWART; and the three firft books have been
tranflated, and repeatedly printed, as an elementary introduction to this branch of fcience. In the preface, Dr.
SIMSON gives a fhort but fatisfactory fketch of the hiftory
of this portion of geometrical fcience from the age of ME-
NÆCHMUS, reputed the firft inventor, to his own time;
in which fome omiffions have been remarked, but not of
much importance in fo fhort an abftract, as it was his purpofe
to communicate.*

To the fecond edition of the *Conics* is added an appendix,
containing two Geometrical Problems, with a preliminary
Locus; which Dr. SIMSON gives as examples of the fuperiority of the ancient analyfis over that of the moderns, in
the refolution of the fame problems, by the authors there
quoted. That fuperiority will be generally admitted; but
it is proper to remark, that befides the folution of one of

* See Note A. at the end.

the problems by GUISNEE,† referred to by Dr. SIMSON, this author, in the fame Treatife, gives another very elegant conftruction of the fame problem, of trifecting the arch of a circle, by means of two local equations.* The Doctor's object, however, was to ftate a comparifon between the pure geometrical analyfis, and the ufual algebraical method, of refolving an equation, and conftructing that folution. The other method by the combination of *Loci,* though clothed in an algebraical drefs, Dr. SIMSON would have confidered, and with reafon, as in effect geometrical; for the *Loci* in queftion might have been eafily deduced geometrically, and thence the proper geometrical folution became obvious.‡

In the year 1752, Dr. SIMSON tranfmitted to the late Earl STANHOPE, in return for fome valuable communications from his Lordfhip, an inveftigation of a rule of approximation to the roots of numbers which are not perfect fquares, given by ALBERT GIRARD in 1629, without demonftration, in his

† GUISNEE *Appl. de Algebre à la Geometrice,* p. 191, 2d ed.

* The fame Problem is refolved with great fimplicity by BOSCOVICH, by means of an Hyperbola, and the circle to which the given arch belongs : fee MAKO *de Arithm. et Geomet. Æquationum Resolutionibus,* p. 332. It may not be improper to mention here a remark by Dr. SIMSON, in a letter to the late Earl STANHOPE, refpecting a Problem propofed by the Doctor's ingenious pupil Mr. WILLIAMSON of Lifbon. " As to the " fubftituting a circle in place of one of the Hyperbolas, I never tried to do it in this " Problem, becaufe I obferved the ancients were not folicitous about fuch folutions, " and preferred the *Loci* which naturally arife from a Problem to any other, as afford- " ing the fhorteft compofition. Thus in the 58th Prop. of the 5th Book of APOLL. " *Conica,* though it is not difficult to folve the Problem by a circle and the given " Parabola, yet APOLLONIUS takes the Hyperbola that arifes from the Problem as " giving the moft natural and fhorteft folution." The letter is dated March 7, 1755.

‡ Dr. SIMSON's opinions on this fubject are more fully ftated in his correfpondence with GEORGE LEWIS SCOTT, Efq; for which fee Note K. at the end.

edition of the works of STEVINUS. That rule feems to have
efcaped the notice of Mathematicians, till Dr. SIMSON undertook the inveftigation of it; which, on the recommendation
of Earl STANHOPE, was fubmitted to the Royal Society, and
publifhed in the Tranfactions of that year. The method
is ingenious, and in fome cafes, may be ufeful; though, from
modern analytical improvements, fuch rules become of inferior confequence.

The next and only other work of Dr. SIMSON which was
publifhed in his life time, was his excellent edition of EUCLID,
which appeared both in Latin and Englifh in 1756, and was
dedicated to his prefent MAJESTY when Prince of Wales. It
contained the firft fix books of the *Elements*, with the eleventh
and twelfth, being thofe ufually taught in Univerfities. From his
annually lecturing on thefe books, and his accurate knowledge
of the ancient geometry, many corrections of the common
text of EUCLID would naturally occur to him; but it was only
after many and repeated folicitations from his mathematical
friends, that he was induced to undertake the preparation of
a new edition. Even after he had nearly compleated it,
a confiderable delay occurred, from doubts which were then
entertained refpecting the conftruction and effect of the
Statute of Queen ANNE, for the encouragement of learning.
He was flattered by his friends with the hope that an edition
corrected by him would have an extenfive circulation, and
would nearly fuperfede the common editions then in ufe;
and he was therefore naturally anxious to fecure the advan-
tages of his work, at leaft for the term prefcribed by the
Statute. But from fome of his letters to the Earl STANHOPE,

it would appear that fuch were the apprehenfions entertained on this fubject, particularly about the edition in Latin, that he had it in contemplation to folicit a fecurity of his right by a private Act of Parliament; and the Act in favour of Mr. BUCKLEY, the Editor of *Thuanus*, was quoted as a precedent.* The Speaker Onslow was confulted; but on full confideration, this fcheme was relinquifhed, and Dr. Simson relied on the protection of the Statute.

To judge with impartiality of the merits of this work, the ftate of the text in preceding editions muft be attended to. Dr. Simson, from his veneration for the ancient Geometers, feems, with an excufable partiality, to have affumed, that the *Elements* of Euclid, as they came from the author, were nearly without blemifh; and he therefore afcribes all the errors and imperfections of the common editions, either to the carelefnefs of tranfcribers, or to the blunders of Theon, and other ancient Editors. His corrections are numerous, and many of them important; and even now, when moft of them are adopted, it might be an ufeful exercife for the young mathematician to ftudy the grounds of his emendations, which exhibit fo clearly the precifion of his ideas, and the logical accuracy of his underftanding. Some animadverfions were made on this edition, chiefly by thofe whofe works had been criticifed in the Doctor's notes; and to fome of thefe, in a fecond edition, replies and explanations were made; but he had a great averfion to controverfy, and his obfervations on what he had

* This Act was for preventing for fourteen years the importation of any Latin copies of that work; which was not provided for in the Statute of Queen Anne.

proved to be errors or defects in his predeceffors, were never
calculated to provoke it.†

Notwithftanding Dr. Simson's valuable corrections, there
are ftill fome difficulties in the *Elements*, which remain to be
cleared up by fome future Editor. The demonftration of the
property of parallel lines (29 I. Elem.) is ftill theoretically
defective, requiring the admiffion of fome principle, not ftrictly
belonging to the clafs of felf-evident truths. It has by fome
been fuppofed, that the remedy for this difficulty muft be
fought for in a juft definition of a ftraight line. No definition
of a ftraight line has yet been found, and none perhaps can
be found, from which all the properties affumed in the
Elements to belong to it, can be rigidly demonftrated. There
is manifeftly alfo fome defect in the definition of a folid angle,
fince what is given in Dr. Simson's, and in all other editions, does
not difcriminate the folid angle from a number of plane angles,
formed at one point, which may exift according to the definition, but without forming the folid angle intended to be
defined. The improvements and corrections of the Fifth Book
are alfo important. His obfervation with refpect to folid figures,
in the note on Def. 10. XI. Elem. is curious, from remarking
an error, which is fo obvious when pointed out, but which had

---

† Soon after the publication of Euclid, Sir Andrew Mitchel, then the Britiſh
Miniſter at the Court of Berlin, aſked Dr. Simson's permiſſion to prefent, in the
Doctor's name, a copy of the Latin Euclid to Frederic III. The Doctor was gratified by the reqneft, and tranfmitted to Sir Andrew Mitchel a copy of the book,
on the page oppofite to the title of which was written the following infcription:

"Cesari in belligerando, Ptolemæis in promovendis artibus et fcientiis, ani-
"mumque fuum doctrina excolendo merito æquiparandus, Magnus Imperator
"FREDERICUS Tertius, Borussorum Rex, ut librum hunc benigne accipiat
"fumma veneratione exoptat. Robertus Simson, Editor."

efcaped the notice of the many learned and acute Geometers, who had paid much attention to EUCLID's *Elements*. An obfervation of a fimilar kind, and about the fame time, was made by Mr. LE SAGE, which is recorded in the Hiftory of the Royal Academy of Sciences at Paris for 1756; and another important correction has been more recently made by LE GENDRE, of which a fatisfactory hiftory is given by Mr. PLAYFAIR, in the fecond edition of his *Elements of Geometry*.

The Book of EUCLID's *Data* was annexed by Dr. SIMSON to a fecond edition of the *Elements* in octavo, (1762,) with many neceffary corrections, and fome valuable additional Propofitions, both of his own and of his learned friends the late Earl STANHOPE, and Dr. STEWART. This book is one of the thirty-three, compofed by the ancient Geometers, for facilitating the refolution of problems by their analyfis; and is the moft proper introduction to the ftudy and practice of that analyfis.* At the end, Dr. SIMSON has added two geometrical Problems, for illuftrating the ufe of fome Propofitions of the *Data*, which would not be obvious to beginners; and with refpect to one of them he obferves, that he believes " it would " be in vain to try to deduce the preceding conftruction from " an algebraical folution of the Problem." The obfervation was perhaps haftily made, when he was feventy-five years of age; but he plainly had in view the moft common method of refolving geometrical Problems by algebra, viz. by deducing a final equation with only one unknown quantity, refolving that equation, and conftructing the folution of it; and with refpect

* In a letter of Dr. SIMSON's, in Note K. are fome obfervations on the ufe of this book.

to this method, the remark would be juft. An ingenious friend of the Doctor's, the late GEORGE LEWIS SCOTT, efq; intimated to him that a conftruction, fimilar to his, might be derived from the combination of two *Loci*, expreffed by two equations arifing in the algebraical folution of the problem. But this communication, with the Doctor's reply, is inferted in a note at the end, which will be more fatisfactory than any abftract of it in this place.*

* See Note K.

SECTION III.

Of Dr. SIMSON's *Pofthumous Works.*

THE ftrong impreffion which Dr. SIMSON felt of the
failure of his memory, having prevented his publifhing
fome important Geometrical Works; the copies of thefe works,
with a large mafs of mifcellaneous papers, fell at his death,
into the hands of his friend and executor Mr. CLOW, the
Profeffor of Logic in the College of Glafgow.†

While Mr. CLOW was deliberating what was moft expedient
to be done with regard to thefe papers entrufted to his care,
the late Earl STANHOPE, diftinguifhed in his elevated rank by
his ingenious cultivation and liberal patronage of the Mathematical Sciences, intimated his defign of publifhing thofe
works of Dr. SIMSON which he had compleated, with any
other pieces, which though unfinifhed, might without injury
to his fame, be given to the public. The munificent propofal
was moft acceptable to Mr. CLOW; and after fome correfpondence refpecting the felection, a large volume, in the
year 1776, was, at his Lordfhip's fole expenfe, handfomely
printed, under the care of Mr. CLOW, and liberally diftributed.

† A confiderable portion of thefe papers confifts of various firft fketches of his
works, publifhed by himfelf, or fince his death.

F 2

This volume contains a reftoration of the *Sectio Determinata* of APOLLONIUS, with two additional books by Dr. SIMSON; and a full explanation of the Porifms, with a reftoration of a number of the Propofitions of EUCLID's original work. To thefe two important works are added fome fmaller Tracts, which fhall be taken notice of in the order in which they are placed in the volume.

The two books of *Sectio Determinata*, it was formerly obferved, had engaged Dr. SIMSON's attention at an early period of his life. That treatife indeed, had been reftored by SNELLIUS, but in a very imperfect manner, without the neceffary diftinction of the various fituations of the Points, called (by APOLLONIUS) *Epitagmata*; and without a compleat expofition of the Determinations, which, as is well known, are neceffary to the perfect folution of any Problem; and are more particularly requifite, as has already been obferved, in the books forming the τόπος ἀναλυομένος, to render them ufeful for the purpofe for which they were compofed and collected.*

In a long and inftructive preface Dr. SIMSON explains the many defects in the Work of SNELLIUS, which need not be here enumerated.† He takes notice alfo of the fubfequent refolutions of fome of the Problems, by ALEXANDER ANDERSONUS;‡ and of thofe likewife contained in the Treatife

---

* See the Preface to Dr. SIMSON's *Reftoration of the Loci Plani*, pp. 7. and 8; where the ufe of the books of the τόπος ἀναλυομένος in refolving Problems is briefly but very clearly ftated.

† It is amufing to obferve SNELLIUS blaming PAPPUS for placing the *Sectio Determinata* after the *Sectio Rationis*, in confequence of his own defective method of reftoring the latter treatife by means of the former.

‡ In *Supplemento* APOLLONII *Redivivi.* Paris, 1612.

of Geometrical Analyfis by Hugo de Omerique; and in the
work itfelf he adopts fome Propofitions from thefe performances.
In the Doctor's firft attempts to reftore the *Sectio Determinata*,
the difficulty was not fo much the giving proper and even
elegant folutions of the various cafes of the general Problem,
as the finding the accurate diftinction of the *Epitagmata*, the
afcertaining the determinations; and chiefly the inveftigation
of folutions, which required the ufe of the *Lemmata* affumed
by Apollonius, and demonftrated by Pappus,* by which
the identity of the reftoration with the original work might be
recognized.   In the preface he obferves, that it was not till
1727,† and after many unfuccefsful attempts, that he obtained
fuch folutions of the fifth ahd fixth Problems of the firft book.
The remaining part of the work required much inveftigation,
and amidft the various other ftudies in which he was engaged,
it was not till above twenty years after, that he compleated
the reftoration to his entire fatisfaction, and in fuch a manner
as to leave no doubt of its being truly the fame as the original
of Apollonius; or fo nearly the fame, as to preclude the
occafion of any further enquiry refpecting it.‡

To the reftoration of the work of Apollonius, two other
books, containing an extenfion of the general Problem, are

* For a more particular account of thefe *Lemmata*, fee the Appendix.

† *Opera Reliqua*, Præfat. ad *Sec. Determ.* p. v.

‡ In a letter to Earl Stanhope, Sept. 10, 1750, he mentions, that for more than
twenty years he had been endeavouring to find the ufe and application of the *Lemmata*
in Pappus, in the folution of the cafes of the *Sectio Determinata*, but had not done fo
completely till laft fummer. And among his papers there is an intimation of his
having juft completed the reftoration of the *Sectio Determinata*, by the ufe of the
Lemmata of Pappus: to which he adds, " a Geometris intacta funt a tempore Pappi
" in hunc ufque diem. 28 Jan. 1749. R. S."

added by Dr. SIMSON, of which it is neceffary for me fhortly to
ftate the hiftory. About two years before his death, when
converfing with him on fome geometrical fubject, he mentioned
that he had compleated the reftoration of the *Sectio Determinata.*
He informed me alfo, that he had written two additional books,
a fair copy of which he took out of his repofitory, and gave
me in confidence, with an injunction to publifh them, or not,
according to the reception which his reftoration of the work of
APOLLONIUS might meet with, among mathematicians. At
that time he was probably thinking of publifhing the work
of APOLLONIUS, but he did not enter into any particular
explanation of his intentions.

. . During his laft illnefs, and for fome time before it, I happened to be at a great diftance from him, and he died without
my receiving any further communication on the fubject. A
fhort time after his death, therefore, I confidered it to be my
duty, after explaining thefe circumftances, to deliver the
manufcript to Mr. CLOW, his particular friend, to whom he had
affigned by his will the property and charge of all his papers.
An account of this fupplement being afterwards communicated
to the late Earl STANHOPE, his Lordfhip was pleafed to defire
that it fhould be annexed to the Doctor's reftoration of the
work of APOLLONIUS.* From the nature of the fubject, and

* At the end of a former copy of the reftoration of this work of APOLLONIUS
is the following. N. B. " Unicuique Problematum quæ in duobus de *Sectione*
Determinata Libris continentur, addenda funt tanquam corollaria in proprijs
" locis, Problemata in quibus quadrata vel rectangula, funt quadratis vel rectangulis
" majora vel minora, dato quam in ratione ; vel quorum unum fimul cum eo quod ad
" alterum datam habet rationum, datum eft. Nam et hæc fæpius in ufu veniunt."
But inftead of this, the Doctor had afterwards thought proper to expand the Problem
in the two additional books juft mentioned.

the conftant references to the two former books, the additional
books are rather uninterefting, efpecially when read without
any particular application of them to the folution of Problems.
But they have a fimilar utility to that of the former books; and
when any geometrical Problems can be reduced to fome particular cafes of this fupplement, it affords alfo an immediate
conftruction and demonftration of fuch Problems.*

The next, and certainly the moft important portion of the
pofthumous volume, contains Dr. SIMSON's difcovery and
illuftration of the Porifms of EUCLID; and as this volume is not
in general circulation, it may be a gratification to fome
readers, befides the account already given, to add a more
particular detail of the hiftory of this curious piece of ancient
fcience, and of the Doctor's fuccefsful inveftigation of it,
after it had baffled the refearches of all the ingenious men
who attempted it before him.

The feventh book of the *Mathematical Collections*, it has
been obferved, is the only notice to be found in ancient
authors of EUCLID's Porifms; if we except the few fhort and
not very fatisfactory obfervations of PROCLUS, who lived, as
is fuppofed, within a century after PAPPUS.† The tranfla-

* Since Dr. SIMSON's death two other reftorations of the *Sectio Determinata* have
been publifhed. One by W. WALES, in 1772; and another, by GIANNINI, for which
fee *Opuscula Mathematica, auctore* PETRO GIANNINI, Parmœ, 1773.—But with
refpect to them, it is fufficient to obferve, that independently of the curious information in Dr. SIMSON's valuable preface, the fuperiority of his work, as a reftoration
of APOLLONIUS, muft be obvious to the geometrical reader.

† PAPPUS flourifhed as is generally reckoned about the year 400 of the Chriftian
Æra, and PROCLUS towards the year 500, and the former is repeatedly quoted by the
latter. See SUIDAS, and GERARD. Voffius *de Univerfæ Mathefeos natura et Confti-
tutione, et Chronologia Mathematicorum*.

tion by COMMANDINE of the fix laft books of PAPPUS,
(which were all that were found in the MS. ufed by him)
was publifhed in 1588.—The attention of the Mathematicians
of Europe was foon directed to that work; and it gave rife
to the various attempts which were made to reftore the loft
works of the ancient Geometers, fo particularly defcribed in
the preface to his feventh book.   The Porifms of courfe
were not neglected; though the very imperfect ftate of the
defcription of them by PAPPUS, in all the manufcripts which
have hitherto been examined, created peculiar, and apparently unconquerable, difficulties in the inveftigation.

ALBERT GIRARD, a Geometer of eminence in the early
part of the 17th century, was the firft who announced the
reftoration of the Porifms of EUCLID.   In his *Trigonometry*
(1629), and alfo in his edition of *Stevinus* (1634,) he ftates
his having reftored the Porifms; but in terms fo general, that
no precife opinion can thence be formed of his notions on
the fubject.   But from the firft of thefe intimations Dr.
SIMSON reafonably infers, that GIRARD was not acquainted
with the true nature of the ancient Porifms; and it appears
alfo to be highly improbable, if he had had any fuccefs in
recovering fo curious a branch of ancient learning, that he
would have concealed it from the world.*

---

* Dr. SIMSON in his preface juftly obferves, that the fhort notices of Porifms in C.
RINALDINUS's work *De Resol. et Comp. Math.* (Bon. 1644) are of no ufe in explaining the ancient Porifms, and in truth have no relation to them.   He feems to ufe
the term Porifm for expreffing the general rule refulting from an algebraical inveftigation in general terms: fee pp. 280, 1.   They are general corollaries, in the common
acceptation of that term.   SCHOOTEN alfo, in his *Exercitationes Math.* (fect. 24.)
ufes the term Porifm for an inveftigation, without any precife object, of a variety of
relations among lines drawn in and about a circle, or other geometrical figure.   He
gives only an example of a circle.

BULLIALDUS is the next author who mentions them in one
of his *Exercitationes Geometricæ*, (1657,) but in it he refers to
FERMAT as the inventor, who had communicated the dis-
covery in letters to fome of his friends at Paris; BULLI-
ALDUS, however, was unable to inveftigate them.* This very
eminent and ingenious man FERMAT appears to have been
the firft in modern times, who made any near approach to
the difcovery of the Porifms. He fuppofed that he had
afcertained the nature of thefe propofitions, and that he
had made fuch progrefs as to enfure a compleat reftoration
of EUCLID's work. In his treatife on the fubject, which was
publifhed only after his death in 1679, he gives five propofitions (without demonftrations) as fpecimens of his invention.
Thefe are certainly Porifms, but none of them belonged to
EUCLID's Treatife ; and this circumftance, befides other arguments, juftified Dr. SIMSON in ftating that FERMAT had
not acquired a correct notion of the nature of a Porifm.†
It is fufficient to remark that FERMAT's definition of Porifms

* Dr. SIMSON, on the margin of his copy of this tract of BULLIALDUS wrote the
following remark: " Ex hoc Tractatu liquet BULLIALDUM nihil de naturâ Po-
" rifmatum intellexiffe. R. S. Mart. 29, 1739."

† An Eloge on Monfieur FERMAT appeared in the *Journal des Scavans*, Feb. 9,
1665, eight years after the publication of BULLIALDUS's tract, in which the fuppofed
difcovery of FERMAT was announced to the world. This Eloge is alfo copied into
the volume of pofthumous works publifhed in 1679 by his fon ; but in the enumeration of his works which it contained, no mention is made of the Porifms. Some
readers may be gratified with FERMAT's definition or defcription of the Porifms.
" Cum Locum inveftigamus Lineam rectam aut curvam inquirimus, nobis tantisper
" ignotum, donec Locum ipfum inveniendæ lineæ defignaverimus, fed cum ex
" fuppofito Loco dato et cognito, alium Locum venamur, novus ifte Locus Porifma
" vocatur ab EUCLIDE, qua ratione Locos ipfos unam fpeciem, et effe et vocari, verif-
" fime PAPPUS fubjunxit."—FERMATII Opera, vol. ii. p. 118.

G

(which at the fame time is not very clearly expreffed,) is avowedly derived from the definition of them by the Mathematicians reckoned modern, (νεώτεροι,) in the time of PAPPUS, and which he pointedly cenfures as inadequate. It may therefore be concluded from this authority, that FERMAT's notion of a Porifm was imperfect; more efpecially as by this definition a numerous clafs of Porifms, altogether unconnected with *Loci*, were neceffarily excluded. The matter, however, is rendered perfectly clear by Dr. SIMSON's full expofition of the nature of Porifms; of which he gives a new definition, very different from that of FERMAT, but fortified by accurate reafoning, and the illuftration of many examples, both of EUCLID's Porifms, and of others compofed by himfelf and his friends. The Doctor alfo obferves, that at an early period he had been ftudying FERMAT's Tract on Porifms, and that in 1715 he had difcovered the demonftration of cas. 1. prop. 4. of FERMAT's Propofitions, before he knew any thing of the nature of Porifms. So little could be derived from the labours of that diftinguifhed Geometrician, by Dr. SIMSON, in his inveftigation of the common object of their purfuit.*

Dr. DAVID GREGORY alfo, an eminent Geometer in the latter part of the fame century, feems, from having taken only a fuperficial view of the fubject, to have deceived himfelf, when he obferves, in the preface to his valuable edition of EUCLID, that it would not be difficult to reftore the Porifms of EUCLID, if the Greek text of PAPPUS were publifhed.

* Dr. HALLEY, in the preface to the *Sectio Rationis*, (laft page) fpeaking of a work of FERMAT's, adds, " qui (fc. FERMATIUS) et Porifmata EUCLIDIS opus longe " difficillimum, redintegrare pollicitus eft, verum fidem non liberavit."

That text was within his reach, as there were two Greek
manufcripts of PAPPUS in the Savilian Library, of which, as
Profeffor of Aftronomy, he had a particular charge; but
there is no evidence of his having made any attempt, and
he died while engaged in another important work, the
publication of the *Conics* of APOLLONIUS, which was finilhed
by his Colleague Dr. HALLEY. From the Savilian manufcripts, Dr. HALLEY was enabled to give an improved edition
of the preface to the feventh book of PAPPUS, with a trans-
lation much fuperior to that of COMMANDINE. The general
account of the Porifms is in fome places corrected ; but the
detail of the contents of EUCLID's work remained fo very
defective and unintelligible, that Dr. HALLEY, with his
great abilities and learning, and with his fuccefsful experience in reftoring and elucidating feveral works of APOL-
LONIUS, was compelled to give up the inveftigation of the
Porifms as hopelefs; and he was brought to this conclufion,
both from the difficulty of the fubject, and alfo from the
mutilations which the injuries of time had occafioned in all the
manufcripts which had been examined.*

From all thefe feemingly unfurmountable obftructions,
it required an uncommon degree of zeal and fortitude in
Dr. SIMSON to undertake the inveftigation; and from his own
account, which has already been given, his perfeverance
was not lefs remarkable. From the time of his firft difcovery
of the nature of a Porifm, the Doctor, without neglecting his
other fcientific purfuits, occafionally applied his mind to this
very curious fubject; and notwithftanding the many difficul-

* See Note B. at the end.

G 2

ties which he encountered, he was at length completely fuccefsful; not only in explaining the diftinctive character of thefe Propofitions, but in the inveftigation of a number of the individual Porifms of EUCLID, and alfo of various other Porifms, which were ufeful, both in illuftrating the nature of thefe Propofitions, and in fhewing the application of them in the refolution of difficult geometrical Problems.

After a certain progrefs in the profecution of this fubject, it became an important object to afcertain a juft definition of the Porifm. The definition given by the later Mathematicians, as ftated by PAPPUS, but cenfured by him, " quod " deficit hypothefi a Theoremate Locali,"* clearly implies that a Porifm had an immediate reference to a Locus ; though it is not lefs clear that PAPPUS confidered Loci as only one clafs of Porifms, (a large one no doubt,) but that of courfe many Porifms have no connection whatever with Loci.

But the definition which PAPPUS quotes from the ancients,† as more characteriftic of Porifms, is too general for any ufeful purpofe; and though it does correfpond to the nature of thefe Propofitions, yet it is deficient in difcrimination, and of itfelf neither conveys any precife notion of EUCLID's Porifms, nor gives affiftance in the inveftigation of any individual Propofition.

* " Quod deeft in hypothefi Theorematis Localis." HALL. verfio.

† PAPPUS, in his defcription of the Porifms, (preface to 7th book,) ftates the following definitions of a Theorem, a Problem, and a Porifm: "Differentias autem horum " trium, melius intellexiffe veteres, manifeftum eft ex definitionibus. Dixerunt enim " Theorema effe quo aliquid proponitur demonftrandum ; Problema quo proponitur " aliquid conftruendum; Porisma vero effe quo aliquid proponitur inveftigandum." HALLII verfio.

After much confideration of various forms of a Definition which had occurred to him, the Doctor finally fettled the following: " Porifma eft Propofitio in qua proponitur demon-
" ftrare rem aliquam, vel plures datas effe, cui, vel quibus, ut et
" cuilibet ex rebus innumeris, non quidem datis, fed quæ ad ea
" quæ data funt eandem habent relationem, convenire often-
" dendum eft affectionem quandam communem in Propofitione
" defcriptam."*

The Doctor illuftrates the propriety and accuracy of this definition by many examples; and fhews particularly wherein the definition blamed by PAPPUS coincides with his, and wherein it is deficient, by excluding many genuine Porifms. The definition indeed, with much addrefs, is fo framed as to correfpond with all the intimations of PAPPUS refpecting Porifms, and alfo with the character of the individual Porifms of EUCLID, which Dr. SIMSON had difcovered; and therefore may juftly be confidered as expreffive of the notions on this fubject entertained by the ancients. It is not pretended that.

---

* From the neceffary generality of expreffion for comprehending every clafs of Porifms, there is fome obfcurity in this definition ; but the Latin language, which Dr SIMSON ufed in all his mathematical writings, is in this cafe better fitted for giving precifion, and for preventing ambiguity, than the Englifh. In Mr. LAWSON's tranflation of Dr. SIMSON's Introduction to the Porifms, which is plainly meant to be ftrictly literal, this definition is expreffed thus: " A Porifm is a Propofition in which it is propofe
" to demonftrate that fome one thing or more things are given, to which, as alfo to
" each of innumerable other things, not indeed given, but which have the fame
" relation to thofe which are given, it is to be fhewn that there belongs fome common
" affection defcribed in the Propofition." The following lefs literal tranflation is
propofed by Mr. PLAYFAIR, for remedying part of the obfcurity: " A Porifm is a
" Propofition in which it is propofed to demonftrate that one or more things are
" given, between which and every one of innumerable other things, not given, but
' affumed according to a given law, a certain relation, defcribed in the Propofition, is-
" to be fhewn to take place." Ed. Tranfact. vol. iii. p. 172, Note.

this was a definition of the ancients; for probably no precife definition was given by them, of either Theorem, Problem, or Porifm. None appears in the works of the more early Geometers, which are ftill preferved in a confiderable degree of purity, and where fuch definitions would naturally have had a place. And we may affirm with much probability, that if any ufeful and characteriftic definition of a Porifm had reached the times of PAPPUS, he would not have neglected fo valuable a remnant of ancient mathematical fcience, in a work obvioufly defigned for the prefervation of the more curious portions of it. He does not omit a definition, which probably was only a traditional and pointed obfervation of fome ancient Geometer; and though of no ufe in explaining the character of a Porifm, yet it in fome degree fortified his objection to the definition of the later Mathematicians, who, he ftates, from inability, could not accomplifh the inveftigation of Porifms; but fatisfied themfelves with affuming the conftructions as they found them in EUCLID, or other Geometers, and adding the demonftrations.

It is obferved by PAPPUS, that a Porifm is neither a problem nor a theorem, but fomething of an intermediate nature; and that it might be propofed either as a problem or as a theorem; fome Geometers contending for the one, and fome for the other. Dr. SIMSON has given a form to the enunciation of a Porifm, implying this intermediate' character between a problem and a theorem. In his enunciation it is affirmed that certain things *may* be found, which fhall have the relations or properties therein defcribed. Perhaps this form refembles more that of a theorem than of a problem; but at the fame time, the things, of which it is faid that they may be found,. muft be actually inveftigated by analyfis, as if the Propofition were a problem. Were it fimply propofed to inveftigate certain things which would have the properties expreffed in the Porifm, it may be regarded as a problem; but if thefe things are found by a conftruction defcribed in the enunciation, the Propofition becomes a theorem, affirming the truth of the properties afferted; and then a demonftration only is required, without any inveftigation; in the manner which appears to have been practifed by the later Mathematicians, alluded to by PAPPUS.*

The moft fatisfactory illuftration of thefe definitions is by examples, of which Dr. SIMSON gives a great variety;

* The enunciation of a Porifm as a problem, is not confiftent with the ufual character of fuch Propofitions. Problems commonly (whatever difficulty attend the actual refolution of them) are almoft immediately recognized by thofe having fome knowledge of geometry, as either poffible in certain circumftances of the data, or as altogether impoffible; and it is unufual to propofe as a problem, " to find things with " certain properties, refpecting the poffibility of which no judgment can be formed " without an analyfis, or fuch confideration as is equivalent to an analyfis." For example, if it had been propofed as a problem in the time of APOLLONIUS, to find in a given parabola a point having the property of the focus, that point being then unknown; fuch a Propofition would not have been confidered as a proper problem, but in reality would have been a genuine Porifm.——PROCLUS, in his defcription of Porifms, mentions the 1 Prop. 3 Elem. " to find the center of a circle," as a Porifm, being in fome meafure between a problem and a theorem. But in that cafe, as a circle, from EUCLID's definition of it, muft have a center, the Propofition " to find that center," feems to be a proper Problem. Had the circle been defined from another of its properties, as for inftance, from its being produced by the extremity of a ftraight line moving at right angles to another ftraight line given in magnitude and pofition, and in the fame plane, fo that the fquare of the moving line be equal to the rectangle by the fegments into which it divides the given line; then the finding of the center would be a proper Porifm; and might be enunciated thus: " within a given " circle (defined in the manner now mentioned) a point may be found, from which " all ftraight lines drawn to the circumference will be equal."

and he diftinguifhes clearly the Locus and Local Theorem
from the Porifm, though all the three, in a large clafs of the
latter, are convertible into each other. He proceeds to give
examples of Porifms altogether unconnected with *Loci*, and
adds likewife an algebraical or arithmetical Porifm; all of
which, however, are clearly comprehended in his own definition, and difplay the defect of the definition cenfured by
PAPPUS.

After an ample expofition of the nature of Porifms, the
Doctor proceeds to the reftoration of fome of EUCLID'S Porifms,
beginning with the general Propofition contained in PAPPUS,
though in an imperfect ftate, and diftributed into the ten
cafes alfo remarked by him; adding the fecond general
Propofition, which is an extenfion of the firft. He then invefti-
gates feveral other Porifms afcertained to belong to EUCLID,
from the remaining fragments of his defcription; and by
employing fome of the *Lemmata* affumed by EUCLID, and
preferved by PAPPUS, he proves the identity of his inveftigation with that of the original author. The Doctor's inveftigation of Porifms, of courfe, is in the ancient analytic method ;
though from the general nature of thefe Propofitions, fome
variation in the form, from what is ufed in a common problem,
is requifite.* The things to be inveftigated, as in the common analyfis, are fuppofed to be found ; and the relation
of them to the innumerable other things muft alfo be affumed,
and often, as exifting in different fituations, to exprefs effectually
the general nature of the Propofition.—It is unneceffary here,
however, to enter more particularly into this difcuffion, as the
propriety and ufe of the form adopted by Dr. SIMSON can

can be fully explained only by examples, and will be intelligible to thofe who have duly confidered what he has
written on the fubject.*

FERMAT's five Propofitions, though ftated by him in
the form of theorems, are undoubtedly Porifms, and as no
inveftigation or demonftration was given of them fo far as Dr.
SIMSON could learn, they are properly added. He gives them
the appropriate form of Porifms, renders them more general,
and adds the neceffary analyfis and compofition. The fecond
Prop. of FERMAT belongs to the parabola, and is demonftrated in the Doctor's *Conics*, Prop. 19. Lib. v. 2d edit.

The Porifms of EUCLID, fo far as may be conjectured from
the defcription of PAPPUS in its prefent imperfect ftate,
feem all to have had reference to the circle and ftraight lines,
and therefore may be called plane; but it is obvious that the
Conic Sections may fupply many folid Porifms, and fuperior
lines in like manner may be the fource of fuperior claffes of
Porifms, which the ancients would have called linear. Dr.
SIMSON concludes this work with fome elegant Porifms of
his own, and fome by Dr. STEWART, one of the very few
to whom he had, in the earlier part of his life, freely communicated his invention.‡ He takes occafion alfo to fhew the

---

* I fhall quote here an important obfervation of Dr. SIMSON's on this fubject.
Opera Reliq. p. 337. " Semel igitur ad minimum, punctum, recta, angulus, vel
" quodcunque fuerit de quo in Porifmate aliquid generaliter affirmatur, fumendum
" eft utcunque, intra tamen limites, fi qui fint in Porifmate præfcriptos, ut legitime
" fiat analyfis. Sæpe autem brevius et facilius ea quæ inveftiganda funt invenientur,
" fi tum quævis ex prædictis rebus fumatur utcunque, tum denuo modo quodam
" particulari fumatur, quando hoc fieri poteft."

‡ See Note, 2 C. at end.

H

utility of Porifms in producing eafy refolutions of geometrical problems, which, without that affiftance, might be complex and difficult.

This application of Porifms by the ancients is ftated by PAPPUS; who infinuates alfo, that little had been done refpecting them by the geometers fubfequent to EUCLID; but as EUCLID's treatife is part of the τόπος ἀναλυομένος, we may prefume that this ufe of the Porifms was in contemplation of the author,† and was probably regarded by fubfequent geometers as the moft important.

I cannot omit adverting in this place to a very ingenious theory of Porifms propofed by Mr. Profeffor PLAYFAIR, of Edinburgh, firft briefly in his account of the life of Dr. STEWART,* and afterwards more fully explained in a memoir on that fubject in the third volume of the Tranfactions of the Royal Society of Edinburgh. The refult of his inveftigation is, that a Porifm is the cafe of a problem which becomes indeterminate; or more particularly, " a Porifm is a Propo- " fition affirming the poffibility of finding fuch conditions " as will render a certain problem indeterminate, or capable " of innumerable folutions."§ But though I admire the ingenuity, and fully admit the foundnefs, of this definition,

† It is proper to obferve that Dr. SIMSON gives a corrected edition of all the *Lemmata* belonging to the Porifms, which muft be very ufeful to thofe who may attempt to recover more of EUCLID's Porifms.

* This life of Dr. STEWART, read before the Edinburgh Society in July 1784, was publifhed in the firft volume of the Tranfactions.

§ In this paper is given a comprehenfive and very ingenious theory of the nature of Porifms founded on this definition; to which the reader who is defirous of fatisfactory information on the fubject is referred.

and alfo the utility of the principle on which it is founded, in the difcovery of Porifms, I muft acknowledge my doubt of that particular notion of a Porifm having ever been adopted, or even propofed, among the ancient geometricians. The circumftance of its being fo fatisfactory as a definition is to me a proof that it was never generally known or embraced : for had it ever been approved and eftablifhed, it feems fcarce poffible that it fhould afterwards have been neglected and loft. That among the ancients the confidera- tion of the relations fubfifting among the data, in fome problems, might have occafionally fuggefted the particular cafe in which thefe problems would become indeterminate, is very probable. It might alfo have often occurred to them, that this indeterminate cafe involved an important general Propofition, which might be feparately ftated as fuch, and preferved. Many Porifms of EUCLID may poffibly have been invented in that way; but ftill I entertain a doubt, if ever the ancients were in poffeffion of this notion as a principle, and as the proper ground of the definition of a Porifm. PAPPUS mentions the definition of the ancients, and apparently as the only one which they were known to poffefs, though, as has been remarked, it be of no particular ufe. He mentions alfo a definition of the later mathematicians, which he cenfures as erroneous : but if fuch a compleat and fatisfactory definition, which not only accurately dif- tinguifhes that clafs of Propofitions, but points out an obvious fource of the difcovery of them, had ever been generally underftood among the ancients, it is difficult to fuppofe that it could ever have been loft; and had it reached the time

of Pappus, it is moft improbable that he fhould neglect
the recording of it in his detailed account of Euclid's treatife
on this fubject. With thefe ftrong internal probabilities,
and the total want of external evidence, I muft (with defer-
ence, however, to the opinion of thofe who may think differ-
ently) adhere to the judgment which I have already expreffed
concerning the recent origin of this excellent definition
propofed by Mr. Playfair.*

The very compleat account of the Porifms given by
Dr. Simson in this pofthumous treatife may be confidered
as fo fatisfactory an illuftration of the nature and character
of Euclid's work, that the juftice of the concluding remark
of his preface will readily be admitted : " His igitur doc-
" trinam Porifmatum fatis explicatam, et in pofterum ab
" oblivione tutam fore, fperare liceat." I may take the
liberty of adding, that though Dr. Simson's fame, as an
accurate, elegant, and ingenious Geometrician, be eftablifhed
from his other works; yet the reftoration of the Porifms of
Euclid will be regarded by pofterity as the moft important
production of thofe powers of inveftigation and genius, with
which he was fo eminently endowed.

There remain fome fmall tracts in the pofthumous volume,
which fhall be very briefly mentioned.

The treatife De Logarithmis is formed on the model of the
fifth book of Euclid; and notwithftanding the great variety

---

* See Note D. at the end. I may here obferve that the two definitions are perfectly confiftent: but though Dr. Simson's might eafily be derived from Mr. Playfair's, yet the deduction of Mr. Playfair's from the other is by no means fo
obvious, and does not appear to have ever occurred to Dr. Simson.

of explanations which have been given of that celebrated and
moft ufeful invention, yet a deduction of the properties of
Logarithms, with the fame ftrict manner of demonftration
which is ufed by EUCLID in treating of proportion, remained
a defideratum in the fcience. It was undertaken at
the requeft of the late Earl STANHOPE, who fuggefted to
him fome valuable hints on the fubject; and the tract, though
fhort, is compleat, and appears to be the laft work which he
finifhed. His correfpondence with the Earl STANHOPE
about the plan of it was in the year 1752; and the fair copy
from which the tract was printed, is dated Feb. 19, 1762. But
in this fome corrections were made with the fanction of the
Earl STANHOPE.

The next tract, *De Limitibus Quantitatum et Rationum*, is
only a fragment; but it contains a rigorous demonftration of
the principles of fluxions, and of prime and ultimate ratios.
It is well known that from the time of the invention of
fluxions, fome inaccurate expreffions refpecting infinitely
little quantities, and fome loofe reafonings founded on fuch
expreffions, had crept into that branch of fcience. Much
difcuffion arofe, and a ferious controverfy was maintained for
fome time; which, in a fcience profeffing to be founded on felf-
evident principles, and conducted with ftrict demonftration, was
of itfelf a matter of fome reproach. Mathematicians of confiderable name took a part in the bufinefs; and many treatifes were
written, with a view to remove the objections which had been
ftarted, and alfo to give a rigid demonftration of the principles
and rules of this new and moft important analyfis. The doctrine of limits was certainly implied (though not in language

calculated to prevent every objection) in Sir IsAAC NEWTON's
Propofitions of prime and ultimate ratios. The principle,
however, came afterwards to be ftated in more accurate
language; and is now by univerfal confent admitted to be the
found foundation on which this branch of the fcience ought
to reft. This certainly has been properly explained in other
treatifes; yet it may be hoped that this fragment of Dr.
SIMSON's will be confidered by the admirers of the fimplicity
and accuracy of ancient geometrical reafoning, not only as free
from even the pretence of objection, but as a demonftration,
with elementary perfpicuity, of the principles of fluxions.*

At the end of the volume is an Appendix, containing a few
geometrical problems refolved in the ancient analytical method.
This addition was made at the requeft of the late Earl
STANHOPE; but it is proper to mention, that it does not appear
that Dr. SIMSON ever intended to make any felection of fuch
problems, with a view to publication. There are, no doubt,
among his papers, a great number of geometrical problems;
fome of which are valuable, but almoft all of them appear to
have been taken up only from the fuggeftion of the moment,
either from his reading, his correfpondence, or from his own
particular ftudies; but generally without any decifive marks of
his having given a particular confideration of the beft poffible
folutions. The few added in this place were felected juft
when the volume was nearly printed, and without any parti-

* It appears from Dr. SIMSON's papers, that he had been thinking of the method
of limits as far back as 1736, and then propofed to extend this method to the firft
and fome other Propofitions of the *Principia*. No trace however remains of his having compleated this defign.

cular object in view, except that examples might be given of the accurate determination of problems, about which the ancient Geometers were fo very curious, and which has often been neglected by the moderns. I muſt obſerve, however, that the laſt problem, being a cafe of the *Tactions* of APOLLO-NIUS, will be mentioned more particularly afterwards.

SECTION IV.

Of Dr. SIMSON's *Unpublished Papers and Correspondence.*

I SHALL conclude the account of Dr. SIMSON's labours and inventions with a few obfervations on his unpublifhed papers, and his correfpondence.* From thefe papers it docs not appear that he ever ferioufly embarked in the reftoration of any other of the treatifes in the τόπος ἀναλυομίνος, except thofe already mentioned. In an early MS. volume, indeed, he gives an arrangement of the cafes of the Problem of the Tactions, and in other papers the folutions of moft of them.†

* Wherever in this Memoir I have made ufe of Dr. SIMSON's unpublifhed papers, the reference to them is particularly remarked, that any inadvertence or miftake of mine might not be attributed to him.

† The Tactions of APOLLONIUS were firft reftored by VIETA, in a treatife, which he calls APOLLONIUS *Gallus.* Afterwards they were reftored by various mathematicians, both geometrically and algebraically. A Treatife of the Tactions by J. GUGL. CAMERER was publifhed at Gotha and Amfterdam in 1795, and is mentioned with commendation by MONTUCLA, (tom. iii. p. 14,) but it contains only an edition of VIETA's treatife, with notes and additions, and a curious hiftory of the Problem. The hiftory is interefting, from the accounts which it contains of the labours of fome foreign mathematicians upon this problem, which are little known in this country. He gives the preface and *Lemmata* of the Tactions in Greek, with fome various readings of feveral manufcripts of PAPPUS. Though VIETA's folutions are elegant, yet they are in feveral refpects deficient. There is not a full diftinction either of the cafes, or of the neceffary determinations. No analyfis is given, and no attempt to reftore the APOLLONIAN folutions by the ufe of the *Lemmata* in PAPPUS, which had been affumed in the work of APOLLONIUS. See CAMERER *de Tactionibus*, p. 4, and p. 12.

It would feem, however, that as there was little difficulty in this problem, it had not been fufficiently interefting to engage him in a compleat reftoration of it. The ufe of one of the *Lemmata*, (Prop. 117, Lib. vii. PAPPI,)‡ in refolving a cafe of the Tactions, did not for fome time occur to him; but having difcovered it, he deduced from it the elegant folution of that cafe which is printed in the pofthumous volume. I found among his papers the following date of this folution: " Feb. 9, ", 1734, mane, poft horam I^{mam} ante meridiem." So minute was he in fome of thefe notices; and in this cafe perhaps, at the moment, he felt a little fatisfaction from having overcome the only difficulty in reftoring the Tactions in the APOLLONIAN method.

PAPPUS, in his defcription of the Tactions, obferves that there are other problems refpecting Tactions, which were generally neglected by the editors; of whom, however, fome prefixed one of thefe problems to the firft of the two books of APOL- LONIUS, as an eafy and proper introduction to the doctrine of Tactions.* PAPPUS gives the general defcription of this problem, from which Dr. SIMSON ftated the arrangement and folution of the feveral cafes; and he alfo propofed and refolved another problem of Tactions, viz. " of points, lines, and " circles, any two being given to defcribe a circle through " the points, or touching the ftraight lines or circles, of which

‡ See Note E. at the end.

* This Problem is " of points, ftraight lines, and circles, given in pofition any two, " to defcribe a circle given in magnitude, which may pafs through the given point " or points, and touch alfo (if poffible) the given lines or line." This problem was " alfo refolved by MARINUS GHETALDUS.

I

" the centre ſhall be in a ſtraight line given in poſition." But all theſe problems, though very eaſy, may be occaſionally uſeful, when a propoſed problem can be reduced to a caſe of any of them.

With reſpect to the *Inclinations* of APOLLONIUS, he ſeems to have beſtowed little attention; only it may be obſerved, that, after repeated unſuccefsful trials, he found (February 20, 1723) the ſolution of the problem about the Rhombus, by the Lemma of PAPPUS, (Prop. 70, lib. vii.) which APOLLO-NIUS no doubt had employed in his ſolution. The figure of this Lemma in COMMANDINE's PAPPUS, which is erroneous, was at the ſame time amended.* The *Inclinations* had been reſtored by MARINUS GHETALDUS ; and ſubſequently to Dr. SIMSON's time, by Dr. HORSLEY, the late Biſhop of St. Aſaph, in 1772; and alſo by Mr. WALES, in 1779. I may further re-mark, that Dr.SIMSON made ſome uſeful corrections of another Lemma of the *Inclinations*, viz. Prop. 85, lib. vii. PAPPI. And in this Propoſition is an example of the deduction of a problem to a caſe of the *Sectio Determinata*, which is thence confidered by PAPPUS as compleatly refolved; and in the demonſtration a reference is made to that caſe, being *Epitagma* 2. Prob. 3. lib. i. *Sect. Determ.* In COMMANDINE's time the loſt treatiſe *De Sect. Determ.* had not been reſtored; and therefore, in his commentary on this Propoſition, he gives a ſolution of that caſe, as neceſſary to the ſolution of this lemmatic problem.

---

* See Note F. at the end, in which is given Dr. SIMSON's ſolution, and the amended figure of the Lemma. It may be obſerved, that an ingenious correction of the Lemma, and of its figure, as it is delineated in the Savil. MS. No. 3, is given by Dr. HORSLEY, in his *Reſtitution of the Inclinations.*

Among the Doctor's papers there remain alfo fome fragments refpecting *Loci ad fuperficiem,* of which EUCLID compofed two books, making a part of the τόπος ἀναλυομένος. PAPPUS, in his preface to his feventh book, gives no defcription of thefe two books, nor any account of their contents, from which, as in other cafes, geometers might have been enabled to explain or reftore them. There are, indeed, a few *Lemmata* at the end of his feventh book, faid there to belong to the *Loci ad fuper-ficiem,* but the firft, as it is enunciated, and the other four,*

* Thefe five *Lemmata* are in COMMANDINE's tranflation, and fome of the MSS. in a very imperfect ftate, as will appear from the following fhort account of them.

The enunciation of Lemma 1, (Prop. 235.) is a fimple Conic *Locus,* but in a fort of fupplementary enunciation is an allufion to a *Locus ad superficiem,* which in its prefent ftate is unintelligible; but a conjecture about the meaning of it is propofed by Dr. WALLIS, in the Savilian copy of COMMANDINE.

In fol. 300. a. COMMAND. 1588, the four laft lines, though in the type of the commentary, is a part of the text, and the Greek, of which it is a verfion, is in the Savil. MS. No. 3. It is the enunciation of a well-known Conic *Locus,* but without any demonftration; and at the end it is faid, " that it will be fhewn by premifing the " following *Locus.*" This probably was Lem. 2, as the next Propofition 236 (viz. the *Locus* referred to) is Lem. 3.

Prop. 236. Lem. 3. is alfo a well-known Conic *Locus,* confifting of three cafes, belong to the three fections; but the firft cafe of the parabola is alone demonftrated in this Propofition.

Prop. 237. Lem. 4. This is the enunciation of the two cafes omitted in the preceding Prop. 236; viz. thofe of the ellipfe and hyperbola, and thefe are analyzed and demonftrated.

Prop. 238. Lem. 5. This propofition is only a repetition of what I fuppofe to have been Lem. 2, and is the fame *Locus* to the three fections as was there ftated. PAPPUS demonftrates only the firft cafe of the parabola, by a reference to Prop. 236; but the other two cafes of the ellipfe and hyperbola are eafily demonftrated by a fimilar reference to Prop. 237.

I may further remark that the *Locus ad hyperbolem,* in Prop. 237, is affumed by PAPPUS in the folution of a folid problem, Prop. 34, b. iv. without however quoting either Prop. 237 of b. vii. or APOLLONIUS. And no doubt it was at the time a well-known folid *Locus.*

I 2

are all folid *Loci*; and truly contain only two *Loci*, of which the one is preliminary to the other. What follows the enunciation of the firft (viz. Prop. 235) in its prefent mutilated ftate is unintelligible; it has, however, fome allufion to a *Locus ad fuperficiem*, but the other four have none whatever. It is indeed fomewhat fingular, that thefe Propofitions are introduced in this place. Pappus, in his preface, profeffes to give an account of the treatifes of the τόπος ἀναλυομένος, only fo far as the *Conics* of Apollonius, in the arrangement affumed by him, which is not the order of time, excluding thereby (from defign, we may fairly fuppofe) the *Loci ad fuperficiem*, and feveral others. Perhaps thefe *Lemmata* may be a fragment of fome other work, as probably is the Propofition, called λῆμμα τῦ ἀναλυομένε τόπε, fubjoined to them in feveral MSS. which has no obvious connection with any other Propofitions of that feventh book. It is poffible, however, that thefe five Propofitions may have been *Lemmata in Locos ad fuperficiem*; but being folid *Loci*, might have been confidered as a proper addition to the " *Lemmata in Conica* Apollonii," to which they are fubjoined.

In the defcription of the *Loci Plani*, however, a fhort account is given of the *Loci ad fuperficiem*, and they are mentioned alfo in Prop. 28 and 29, lib. iv. Pappi; which are problems refolved by means of *Loci ad fuperficiem*, and which Dr. Simson has corrected and rendered intelligible. Thefe are all the materials remaining for the inveftigation of this difficult and curious, though perhaps not very important, piece of ancient geometry.

Dr. SIMSON had begun very early to confider thefe *Loci*;
for in 1721 I find fome equations for expreffing the relation of
the co-ordinates, by means of three indeterminate quantities.
It appears alfo that he had afterwards defigned to give a regular
explanation of thefe *Loci*; for there remains a preface for fuch
an effay, in which he propofes to explain the method of inveftigating fuoh *Loci* by examples... From thefe he hopes that
it would be eafy for one well acquainted with folid *Loci* to
inveftigate and defcribe any *Loci ad fuperficiem* which can
arife from Conic Sections varioufly fituated; (" *vario fitu
difpofitis.*") In the beginning of the preface, where he remarks
the very little known of thefe *Loci* among the moderns, he has
a marginal correction, intimating that he had not, at the time
of writing this preface, feen the treatife *Des Courbes a double
Corbure*, which he afcribes (furely from inadvertence) to D. N.*
But as there is no treatife of that period of this title, or indeed
on this fubject, but that of M. CLAIRAUT, publifhed in 1730,
there can be little doubt of that being the work alluded to by
Dr. SIMSON.†

* M. NICOLE, in a Memoir of the Academy of Sciences, (1734) but fubfequent to the
publication of CLAIRAUT's work, makes an allufion to the curves of double cur-
vature. M. NICOLE was one of the cenfors of CLAIRAUT's treatife.

† The very flight notice of the ancient *Loci ad superficiem*, by DES CARTES, in his
Geometry, (book ii.) was of no ufe for Dr. SIMSON's inveftigation. " At vero duabus
" conditionibus deficientibus ad hujus puncti determinationem, Locus in quo illud
" reperitur fuperficies eft," &c. This, to be fure, correfponds to an equation with
three unknown quantities, but it is not purfued by DES CARTES. His commentator
SCHOOTEN gives two examples, which may perhaps be included in the general
defcription of the *Loci ad superficiem* by PAPPUS, in his account of *Loci Plani*; but
the Data and the Locus are all in the fame plane, and they furnifh no illuftration of
what Dr. SIMSON confiders as the object of the ancient treatife of EUCLID.——See
CARTESII *Geomet.* Amft. 1683, p. 34; and SCHOOT. *Comment.* p. 228. See alfo
SCHOOTENII *Sectiones Miscellaneæ.* Lugd. Bat. 1657, p. 494.

All thefe are comprehended in the following equation:

$$xx + \frac{a}{b}xy + ax + by + \frac{a}{c}yy + \frac{a}{d}yz + \frac{a}{e}xz + gz + \frac{b}{k}zz = \text{ Dato.}$$

The figns being any how changed.*

In proceeding however to examples, he treats the fubject geometrically, and without any reference to the equation. There are indeed feveral examples, but all of them, except a few of the fimpleft forms, are unfinifhed; and though in the firft there is a confiderable progrefs, and an inveftigation of the common fection of the *Locus ad fuperficiem* fought, when cut by feveral planes in different pofitions, yet he mentions feveral others to be afcertained, but which are not further profecuted. The Doctor was not eafily diverted from an inveftigation, merely from the difficulty of it; and we may therefore reafonably infer, that he abandoned the fubject, partly from its having been previoufly taken up by CLAIRAUT, and partly from confidering it as a branch of ancient geometry, of inferior importance and utility.† The omiffion of any defcription of this and fome other treatifes of the τύπος ἀναλυομένος, in the preface to the feventh book, is a prefumption of a fimilar eftimation of the *Loci ad fuper-ficiem* being entertained by PAPPUS.‡

---

* See Note G.

† See Note G. at the end, fome of the eafieft examples.

‡ In vol. ii. p. 151, of FERMAT's works, there is a letter of his to ROBERVAL, in which, after fome obfervations about *Loci*, and particularly refpecting one of the fecond book of APOLLONIUS, he adds, " *de Locis folidis et ad fuperficiem multa quoque* " *funt detecta*." He flightly alludes to the fame fubject in a letter to Sir KENELM DIGBY. WALLISII *Opera*, tom. ii. p. 859. Alfo in page 116 of fecond vol. of FERMAT's works, in the preface to his Porifms, after mentioning the reftoration of ancient treatifes of geometry by SNELLIUS, VIETA, and MARINUS GHETALDUS, he adds, " fequebantur *Loci plani, Loci folidi*, et *Loci ad fuperficiem*, et huic quoque

Befides thefe fragments of geometrical fpeculations, there
are a great variety of Problems, Theorems, Loci, Data, and
Porifms, difperfed through the Doctor's papers, and particularly in the MS. volumes, which he called *Adversaria*, of which
eighteen remain. Thefe in general, as has already been
obferved, appear to have been analyzed or demonftrated,
as the circumftances of his reading, converfation, or ftudy,
happened to fuggeft; yet they may be confidered as a valuable
repofitory of fuch Propofitions, from which ufeful felections
might be made, for promoting the ftudy of geometry in the
ancient method. They are depofited in the College library
of Glafgow; and it is much to be wifhed, that fome gentleman
acquainted with the fubject, and who has leifure and inclination for the tafk, would take the trouble of arranging a
proper collection.*

It has been repeatedly remarked, that notwithftanding Dr.
SIMSON's early and continued partiality for geometry as

" parti non ignoti nominis geometræ, fuccurrerant, eorumque opera manufcripta,
" licet et adhuc inedita, latere non potuerint." And in another letter to Mr.
ROBERVAL (p. 153) he obferves, " J'en ay plus de cens Propofitions tres belles et
" particulierement des *Lieux solides* et *ad superficiem*, mais le loifir me manque, &c."
No further account however of thefe reftorations of *Loci ad superficiem* is to be
found.

* There are in thefe volumes, alfo, a number of geometrical demonftrations, by the
Doctor, of theorems, in the works of various eminent analyfts, where they are treated
either algebraically, or if geometrically,with inferior fkill.—I may remark that moft of
the Propofitions alluded to are dated; and on fome particular occafions not only the
day, but even the hour is noted. Sometimes alfo, when he had accomplifhed a
folution or a demonftration, which pleafed at the moment, fome expreffion of his
fatisfaction is added. He began this practice very early, even before he was elected
Profeffor of Mathematics, and it was continued till about fourteen months before his
death. In the book of *Adversaria* Q. is the lateft date now to be found, 11th Auguft
1767; he being then in his eightieth year.

cultivated by the ancients, he did not entirely neglect the
modern analyſis, but became well acquainted with algebra,
and with the common applications of it which were purſued
in the early part of the laſt century.* He certainly did not
approve of the prevailing uſe of algebra in geometrical inveſtigations, which could be accompliſhed with more elegance,
and not ſeldom with more eaſe, by the ancient method ; but
he made no ſcruple of acknowledging the ſuperior power
of algebra in complex geometrical inquiries, and more eſpecially in the application of it to ſuch as are phyſical.—It is
probable however, from all the information which can now be
obtained reſpecting his ſtudies, that though he underſtood the
modern analyſis as delivered by Sir ISAAC NEWTON, yet he
did not continue to devote much of his time to the ſtudy of
it, and thence did not enter much into the profound analytical
ſpeculations of ſucceeding mathematicians.

It muſt be admitted, alſo, that his rigid notions reſpecting
algebraical reaſoning, particularly in geometry, had a tendency
to limit inveſtigations, by that method ; and to diſcourage him
from purſuing the methods of modern analyſts, who felt no ſuch
difficulties as had occurred to him. As an example, however,
of Dr. SIMSON's knowledge of algebra, I may mention his invention of ſome very ſimple and excellent ſerieſes for finding the
circumference of a circle, of which he appears to have given
an intimation in 1722 to Dr. JURIN, then ſecretary of the
Royal Society.—Dr. JURIN undertook to obtain the ſentiments

---

* Dr. SIMSON often amuſed himſelf, eſpecially in the latter part of his life, in
reſolving any geometrical problems which attracted his notice, by the algebraical
method, though he rarely made any entry of theſe ſolutions in his *Adverſaria.*

of his learned friends on thefe feriefes; and therefore Dr.
SIMSON, in a letter of Feb. 1, 1723, tranfmitted to him his
paper on the fubject,* containing the feriefes, with two
elementary Propofitions from which they were derived.
Though Dr. JURIN was not apprized that any thing of the
kind had been done by others, yet by inquiry he found that
Mr. MACHIN, near twenty years before, had communicated
to the Royal Society fome feriefes of the fame form, but had
afterwards withdrawn them. Mr. MACHIN readily gave a
copy of his feriefes to be communicated to Dr. SIMSON,
which Dr. JURIN fent to him immediately after; (March

---

* I annex a copy of the portion of this letter which relates to the feriefes. The
remainder of it refpects the Porifms, and is taken notice of in another place. For the
paper enclofed, fee Note H. at the end.

   " SIR,
   " Friday laft I had the favour of your very civil anfwer to the letter I lately gave
" you the trouble of: fome private affairs have kept me from transfcribing the
" enclofed trifle, elfe you fhould have had it fooner.—The feries for finding the
" circumference of the circle which are in Mr. SHERWIN's Book of Tables, being
" deduced from tangents of arches which have a given ratio to the circumference, I
" was defirous to try if the like might not in a general way be drawn from tangents,
" whofe arches are not in a given ratio to it; and found a great many eafily flowed
" from the two Propofitions and their corollaries, which I have enclofed. The Pro-
" pofitions, though obvious enough, I do not find in Mr. BRIGGS, or any of the authors
" I have fearched for them: the corollaries are long fince known, if I miftake not.
" Among the feries I have fet down, there is one which Mr. JONES, in his Synopfis,
" fays he had from Mr. MACHIN, and which is certainly one of the beft that can be
" found by this or any other method. Whether any or all of the reft have been taken
" notice of by any body, I am wholly ignorant; but you may foon difcover.
   " I thankfully acknowledge the favour of your kind offer to obtain the fentiments
" of fome of your learned friends upon the enclofed paper, but I entreat you may
" give yourfelf no more trouble about it, fince I am afhamed to have given you
" already fo much, when the thing is fcarce worth taking notice of, and which you
" will eafily fee is not to be fhewn to any but as fuch."
       " Glafgow, Feb. 1, 1723."

K

1723). The extreme modesty with which Dr. SIMSON
mentions this ingenious inveſtigation to Dr. JURIN, is remarkable; for though from the preſent ſtate of ſcience, thoſe
ſerieſes may have loſt their former importance, yet near a
century ago they had the merit of novelty and utility, and
had then occupied the attention of Mr. MACHIN and Dr.
SIMSON. It may thence be ſatisfactory to ſome readers to ſee
a fuller account of them, which is given in a note at the end.*

The Mathematicians of the end of the ſixteenth and
beginning of the ſeventeenth centuries began to apply
algebra to geometrical inquiries; and this method was much
promoted and extended by DES CARTES. The uſe of
negative and impoſſible roots of equations was alſo introduced
into this analyſis; though by preceding algebraiſts they
were at firſt not obſerved, and afterwards when obſerved were
reckoned uſeleſs. While algebra, in its early and humble
ſtate, was employed almoſt entirely about numbers, theſe
roots were naturally neglected, and the language of the firſt
writers on the ſcience was thence not encumbered with the
metaphyſical difficulties which aroſe from the proceſſes and
reaſonings of the followers of DES CARTES. Some of theſe
were reaſonably conſidered by Dr. SIMSON as defective in
that preciſion of definition, and ſtrictneſs of argument, which
have ever been the boaſt of pure geometry, and to which he
had been accuſtomed from his almoſt conſtant ſtudy of that
branch of the ſcience. The Doctor, from this cauſe perhaps,
conceived a prejudice againſt an application of algebra, which

* See Note H.

was accompanied with fuch difficulties; and was thence led to treat of that fcience after the manner of the early writers on it, with the limited definition and ufe of the negative fign. Many detached obfervations on this fubject remain among his papers;* and fome fhort effays alfo on cubic equations, in which he endeavours to explain them without admitting negative or impoffible roots. In the prefent ftate of that fcience, it is unneceffary to mention the detail of thefe fpeculations; and it may be fufficient to remark, that though the method of inveftigation was different, the refults were fimilar to thofe contained in the Effay on the Negative Sign by that learned promoter and patron of mathematical fcience, Mr. BARON MASERES; of which, from the correfpondence of opinions, Dr. SIMSON fpoke with much approbation. Dr. SIMSON's notions on this fubject probably varied a little in the progrefs of life, by his confirmed partiality for the ancient analyfis, and an encreafing prejudice againft the algebraical. The elegance and fatisfactory clearnefs of every ftep in the geometrical method muft, where it can be employed, be univerfally preferred to the almoft mechanical procefs of an algebraical calculation. His animadverfions, however, on the application of algebra to geometry, chiefly referred to thofe cafes where it was not neceffary, and in which the more excellent method of the ancients could be fuccefsfully employed. In Plane Geometry, and in the Conic Sections, the ancient method is ftill generally ufed, and by good judges is allowed to be the beft for forming the

* See in Note K. at the end, fome obfervations by Dr. SIMSON on this fubject, in his letter to Mr. GEO. LEWIS SCOTT.

tafte, and exercifing the reafoning powers, of young perfons; and Dr. SIMSON has done effential fervice by his geometrical works, in exciting curiofity refpecting thefe branches, and facilitating the ftudy of them.

Dr. SIMSON has no where in his writings given any fpecific opinion of the proper and neceffary ufe of algebra in geometrical enquiries, and therefore we cannot judge with precifion of his views refpecting it. It may be obferved, however, that as the indefinite number of claffes of geometrical curves, in the prefent ftate of both the ancient and modern analyfis, can be conveniently treated of only by algebra, this application of algebra muft be admitted as neceffary ; and the definition of the Conic Sections by equations may be regarded as an ufeful introduction to the definition of fuperior curves on that principle. Dr. SIMSON feems, in the early part of his life at leaft, in fome meafure to have admitted this; for in a great number of folid *Loci* among his papers, though demonftrated geometrically, the fuitable equation is added. Probably, however, he never fet about taking an accurate and comprehenfive view of the fubject in all its bearings ; and having made his early choice from tafte, the ancient analyfis continued through life to be the favourite object of his ftudy.

Confidering the unbounded extent of mathematical fcience, and the variety of branches into which it may, and has been ufually diftributed ; it may be for the advantage of the moft fuccefsful progrefs of the whole, that individual mathematicians fhould apply themfelves chiefly to particular divifions of it. Though there be a clofe connection among

thefe feveral branches, yet a moderate knowledge of the
general fyftem may be fufficient for the fuccefsful cultivation
of any one portion of it, to which the attention of a mathematician may happen to be directed. Had Dr. SIMSON applied.
his ardent mind to the modern analyfis, moft probably the
refults of his inveftigations would have been important; but his
not having devoted much of his attention to the ftudy of it, was
certainly favourable to his progrefs in the laborious refearches
which he undertook refpecting the ancient geometry. It will
be readily admitted alfo, that if fome of the great modern
analyfts happen not to have ftudied the ancient geometry, as
may be prefumed to be the cafe, they were not thereby
materially obftructed in their profound inveftigations, by which
human knowledge has been fo much enlarged; though,
perhaps, by fuch previous ftudy, their works might in fome
inftances have been improved in fimplicity and perfpicuity.

It appears from the Doctor's letters to the late Earl STAN-
HOPE,* that his Lordfhip had recommended to him to attempt
the demonftration of fome of the more important modern
geometrical difcoveries by the ftrict and more elegant method
of the ancients. This was in 1751, and though he ftates it to
be very much his own wifh to make the trial, he confiders it
to be beyond his power, from the decline of his faculties. The
tract on *Logarithms*, and the fragment *De Limitibus*, feem to
be his only attempts of that kind ; but he propofes to recommend it to his fcholars, Dr. STEWART, Dr. MOOR, and Mr.
WILLIAMSON. There are indeed among his papers a few
imperfect fketches (apparently of an old date) of attempts to

---

* See Note I. at the end.

demonftrate geometrically the celebrated theorem of the Circle invented by Mr. Cotes.   He does not feem to have beftowed much time to this fubject; but I have heard him mention that his plan was to demonftrate firft the more eafy cafes, and then to attempt to frame a general method of deducing any cafe of this Propofition from the next preceding and more fimple cafe, from which a compleat demonftration of the theorem in its full extent might be derived.   But in this he was unfuc-cefsful, and he regretted the difappointment.   But as neither Dr. Simson, nor any other geometer, fo far as is known, have ever accomplifhed a geometrical demonftration of this theorem, we may confider the inveftigation of Mr. Cotes as an important example of the fuperior power of the modern analyfis, in the difcovery and demonftration of a Propofition, very general and difficult indeed, but ftrictly geometrical; and though it be commonly expreffed algebraically, yet may, by means of compound ratios, be enunciated in the pure language of geometry.

Long before any fymptom of decline in his health appeared, he feems to have felt fome remarkable decay of his memory and attention.   This was not obferved by his friends, either in converfation or in correfpondence; but his complaints of it were fo particular and ferious, that no doubt can be entertained of the fact, nor of the influence which it had in diverting his mind from fome of the more fevere inveftigations which he had formerly wifhed to purfue.*

In the preface to the Doctor's *Conic Sections,* when recommending the geometrical method of treating the fubject, in

* See Note K. at the end.

preference to the algebraical, he has this remarkable sentence :
" In quibus autem differat analyfis geometrica ab eâ quæ
" calculo inftituitur algebraica, atque ubi hæc aut illa funt
" ufurpanda, atque quæ fint in mathematicis utriufque partes
" propriæ, alias differendum." From this obfervation it muft
be acknowledged, there naturally arifes a prefumption, which
has been generally entertained, that the Doctor had confidered
the fubject, and probably written fomething concerning it:
and the opinion of fo diftinguifhed a mathematician on a point
of fome difficulty, and of which he was well qualified to judge,
would have been particularly interefting. But among his
papers no trace of fuch an attempt is to be found. Profeffor
Robison, in his Life of Dr. Simson already mentioned, feems
to think that a differtation on *Loci,* which he had feen when
attending the Doctor's Lectures in the College of Glafgow,
might have been the tract to which the preceding quotation
referred.* Of this, however, notwithftanding the weight of
Mr. Robison's opinion, a reafonable doubt may be entertained, confidering the extent of the difcuffion implied in that
quotation, which muft have gone far beyond what fuch a differtation as Mr. Robison mentions, could be fuppofed to
contain. But Dr. Simson's own teftimony, in his correfpondence with Mr. Scott refpecting this very fentence, is con-
clufive, and puts an end to all fpeculation regarding it, as he
therein explicitly declares, that on the fubject of it " he had no
" papers, and never wrote any thing on that matter." Fortunately, a confiderable part of this correfpondence is preferved ;
and the moft important portion of it is placed in a note at the

* No trace of this differtation on *Loci* remained among his papers.

end.* The letters both of Mr. Scott and Dr. Simson will be read with intereft: they contain fome valuable obfervations on the relative merits of the ancient analyfis and of the algebraical, in the folution of geometrical problems. The fubject indeed is not exhaufted; and as the correfpondence took place from occafional circumftances, the difcuffion, though entertaining and inftructive, is of courfe limited very much to the points which thus accidentally arofe. It touches little on the peculiar powers of the modern analyfis, and is confined chiefly to the comparifon of the two methods in cafes where either may be employed.†

* See Note K.

† For fome remarks on this correfpondence, fee the latter part of Note K.

SECTION V.

Sketch of Dr. SIMSON's *Character.*

IT was formerly obferved that Dr. SIMSON's mathematical labours and inventions, of which fome account has been given in this memoir, were the moft interefting occurrences of his life, and moft illuftrative of his capacity and genius.

In attempting, however, to give a brief outline of the Doctor's character, befides the talents difplayed in his geo‑metrical works, which claim the firft confideration, it may be expected that fome obfervation alfo fhould be made of the well-known difcriminative features of his mind, which, though not important in themfelves, appear to have had an influence on its greater qualities. Thefe peculiar features are ftill affectionately remembered by his pupils, friends, and acquaint‑ances who furvive; and, from the eminence of his genius, they may be regarded by others as objects of a natural and liberal curiofity.

Dr. SIMSON was originally poffeffed of great intellectual powers, an accurate and diftinguifhing underftanding, an inventive genius, and a retentive memory: and thefe powers, being excited by an ardent curiofity, produced a fingular capacity for inveftigating the truths of mathematical fcience. By fuch talents, with a correct tafte, formed by the ftudy of

L

the Greek Geometers, he was alfo peculiarly qualified for communicating his knowledge, both in his lectures and in his writings, with perfpicuity and elegance. He was at the fame time modeft and unaffuming; and though not indifferent to literary fame, he was cautious and even referved in bringing forward his own difcoveries, but always ready to do juftice to the merits and inventions of others. Though his powers of inveftigation in the early part of life were admirable, yet before there appeared any decline of his health, he felt ftrong impreffions of the decay both of his memory and other faculties; occafioned probably by the continued exertion of his mind in thofe fevere ftudies, which for a number of years he purfued with unremitting ardour.

Befides his mathematical attainments, from his liberal education he acquired a confiderable knowledge of other fciences, which he preferved through life, by occafional reading, and in fome degree alfo, by his conftant intercourfe with fo many literary men in his College.† He was efteemed a good claffical fcholar, and though the fimplicity of geometrical demonftration does not admit of much variety

---

† Dr. Simson in the early part of his life ftudied Botany, in which he made great progrefs, and it became a fource of amufement to him in his walks. In the year 1746, when the Univerfity of St. Andrew's wiſhed to confer on him the degree of Doctor, as he was a layman, a degree of medicine was propofed from the circumftance juft mentioned, though he had no other pretenfion to diftinction in that fcience.

It may be mentioned alfo in this place, that a few years after he became a Profeffor, he compofed a tract on the School Logic, with rigorous demonftrations of its rules. Some fketches of it remain, which ftrongly mark the accuracy of his reafoning powers to whatever fubject they were directed.

of ftile, yet in his works a good tafte in that point may be diftinguifhed. In his Latin prefaces alfo, in which there is fome hiftory and difcuffion, the purity óf language has been generally approved. It is to be regretted indeed that he had not had an opportunitity of employing, in early life, his mathematical and Greek learning, in giving an edition of PAPPUS in the original language.‡

Dr. SIMSON never was married; and the uniform regularity of a long life, fpent within the walls of his College, naturally produced fixed and peculiar habits, which however, with the fincerity of his manners, were unoffending, and became even interefting to thofe with whom he lived. The ftrictnefs of thefe habits, which indeed pervaded all his occupations, probably had an influence alfo on the direction and fuccefs of fome of his fcientific purfuits. His hours of ftudy, of amufement, and of exercife, were all regulated with uniform precifion. The walks even in the fquares or garden of the College were all meafured by his fteps, and he took his exercife by the hundreds of paces, according to his time or inclination.

‡ Dr. SIMSON's learned difcuffion concerning fome Greek terms in the account of the Porifms by PAPPUS has already been mentioned, and a copy of it is inferted in Note C. I may mention here alfo another example of his critical knowledge. In the dialogue of PLATO, MENO, there is a paffage, of which, even in the beft editions, fome fentences are fomewhat obfcure, and probably corrupted. It relates to an ingenious fpeculation, of all knowledge being only reminifcence; and as an illuftration, an uninftructed youth is led, merely by an examination, to affent to the truth of a geo-metrical propofition, (an eafy cafe of the 47. 1. ELEM. EUCL.) This had attracted Dr. SIMSON's notice; and he found that fome natural and eafy emendations, with the addition of a diagram, which, in the dialogue, is plainly fuppofed to be exhibited to the youth, rendered the argument intelligible and correct. He had drawn up a detailed ftatement of the paffage, and the neceffary corrections which, in his own time I faw in the hands of one of his colleagues, who has been dead many years: but no trace of it is now to be found among the Doctor's papers, or any where elfe.

It has been repeatedly mentioned, that an ardent curiofity was an important feature of his character. It contributed effentially to his fuccefs in his mathematical inveftigations, and it difplayed itfelf in the fmall and even trifling occurrences of common life. Almoft every object and occurrence excited it, and fuggefted fome problem which he was impatient to refolve. This difpofition, when oppofed, as it often neceffarily was, to his natural modefty, and to the formal civility of his manners, occafionally produced an embarraffment, which was amufing to his friends, and fometimes a little diftreffing to himfelf.

The Doctor, in his difpofition, was both cheerful and focial; and his converfation, when he was at eafe among his friends, was animated and various, enriched with much anecdote, efpecially of the literary kind, but always unaffected. It was enlivened alfo by a certain degree of natural humour; and even the flight fits of abfence to which in company he was occafionally liable, contributed to the entertainment of his friends, without dimi-nifhing their affection and refpect, which his excellent qualities were calculated to infpire. One evening in the week he devoted to a club, chiefly of his own felection, which met in a tavern near the College. The firft part of the evening was employed in playing the game of whift, of which he was particularly fond; but though he took no fmall trouble in eftimating chances, it was remarked that he was often unfuccefsful. The reft of the evening was fpent in cheerful converfation, and as he had fome tafte for mufic, he did not fcruple to amufe his party with a fong; and it is faid that he was rather fond of finging fome Greek odes, to which modern mufic had been adapted. On Saturdays he ufually dined in the village of

Anderſton, then about a mile diſtant from Glaſgow, with ſome
of the members of his regular club, and with a variety of other
refpeƈtable viſitors, who wiſhed to cultivate the acquaintance,
and enjoy the fociety, of fo eminent a perfon. In the progrefs
of time, from his age and charaſter, it became the wiſh of his
company that every thing in thefe meetings ſhould be direƈted
by him; and though his authority, growing with his years,
was fomewhat abfolute, yet the good-humour with which it
was adminiftered, rendered it pleaſing to every body. He had
his own chair and place at table; he gave inſtruƈtions about the
entertainment, regulated the time of breaking up, and adjuſted
the expenfe. Thefe parties, in the years of his fevere ftudy,
were a defirable and ufeful relaxation to his mind, and they
continued to amufe him till within a few months of his death.

Striƈt integrity and private worth, with correfponding purity
of morals, gave the higheſt value to a charaƈter, which, from
other qualities and attainments, was much refpeƈted and
eſteemed. Upon all occaſions, even in the gayeſt hours of
focial intercourfe, the Doƈtor maintained a conſtant attention
to propriety. He had ferious and juſt impreſſions of religion,
which appear occaſionally in his papers, and may be traced
even in the midſt of fome of his mathematical inveſtigations;*
but he was uniformly referved in expreſſing particular opinions
on the fubjeƈt of religion; and from his fentiments of decorum,

* As an example, the following date of the folution of a problem, which happened
to be his birth-day, may be quoted, as it ſtands in one of his MS. volumes:
                "14 Oƈtobris 1764
                "14 Oƈtobris 1687
                        ——
                        "77
    "DEO Optimo Maximo, Benigniſſimo Servatori, fit laus et gloria."

he never introduced religion as a fubject of converfation in mixed fociety, and all attempts to do fo in his clubs were checked with gravity and decifion.

Dr. Simson in his perfon was tall and erect, and his countenance, which was handfome, conveyed a pleafing expreffion of the fuperior character of his mind.* His manner had always fomewhat of the fafhion which prevailed in the early part of his life, but was uncommonly graceful. He was ferioufly indifpofed only for a few weeks before his death, and through a very long life had enjoyed an uninterrupted ftate of good health. He died on the firft of October 1768, when his eighty-firft year was almoft compleated; having bequeathed his fmall paternal eftate in Ayrfhire to the eldeft fon of his next brother, in whofe family it ftill remains.†

* In the College hall of Glafgow is a portrait of Dr. Simson, which, though painted when he was in the vigour of life, yet expreffes a likenefs to his countenance and figure as they appeared in advanced age. On the title-page of this Memoir is placed a fmall engraving from this picture, with an infcription written by the late Dr. Moor, Greek Profeffor at Glafgow, happily expreffing Dr. Simson's difapprobation of the modern ufe of algebra in geometry, and his fingular merit and fuccefs in reftoring the ancient analyfis. See the Edinburgh Tranfactions, vol. i. p. 75. Hiftory.

† Dr. Simson, from his conftant refidence in the College, and his engagement in ftudy, gave little attention to the improvement of his eftate, which defcended to him encumbered with debts, and which became much more productive to his heir than it had been to himfelf. He was at the fame time well acquainted with bufinefs, having been for a great number of years Clerk to the Faculty, and in that character he had the chief management of the property and other affairs of the College, which he conducted with his habitual regularity and accuracy. In his manner of living he was fimple and unexpenfive, except in collecting a valuable mathematical library, which he bequeathed to his College. But from an income, ample with refpect to his habits, he made no accumulation; having enjoyed the fatisfaction of advancing young relations, in his own time, by many acts of generofity.

NOTES.

NOTES.

NOTE A. p. 28.

FROM APOLLONIUS and his commentators to MYDORGIUS and others of the feventeenth century, Dr. SIMSON confiders this branch of geometry as ftationary for that long period. It has been properly obferved however by different authors, that the points, in fubfequent times called *Foci*, were afcertained by APPOLLONIUS only in the ellipfe and hyberbola, but not in the parabola. The Rev. Dr. ROBERTSON, Savilian Profeffor of Aftronomy at Oxford, in his learned Hiftory of Conic Sections, annexed to his Treatife on that fubject,* after mentioning this obfervation, adds, that in two fmall tracts, now very rare, edited by GOGAVA, Louv. 1548, the firft mention is made, fo far as he knew, of the focus of the parabola.† GOGAVA confiders the fecond tract (*De Speculo Ustorio*) as of Arabic origin, and very ancient; and alfo that VITELLO had borrowed from it. VITELLO flourifhed about 1470,

* *Sectionum Conicarum, Libri VII.* &c. Oxon. 1792, pp. 340, 362.

† This publication appears to be a firft edition, though other dates are given to it by writers who quote it; and it has a Preface by GEMMA FRISIUS. The fecond of thefe two fmall tracts feems to have been generally known before the edition by GOGAVA. ROGER BACON quotes it; and VERNERUS, in his tract *De Cubo Duplic.* Nurem. 1522, append. 12mo. plainly alludes to it. MAUROLYCUS alfo, in a mifcellaneous volume, (Meffanæ, 1558,) ftates

M

near a century before GOGAVA; and in his *Optics*, prop. 42, 43, book ix. (page 400, edit. Refneri,) the focus of the parabola, and the application of it in forming a concave parabolic mirror are diftinctly defcribed, but without ftating the fource from which he derived them.

It is plain, however, that the point, afterwards called the focus of the parabola, was ufed by PAPPUS, though not formally diftinguifhed or named, in prop. 238, lib. vii. (fol. 303. b. edit. 1588,) which is a *Locus* to the parabola. But he does not mention the property, eafily flowing from that which he affumes, on which is founded the conftruction of the parabolic reflecting mirror; and it is not improbable that this laft property had not been obferved in the time of PAPPUS, and that it was difcovered long after by the Arabian opticians. This Prop. of PAPPUS muft have been well known to Dr. SIMSON, though he appears not to have confidered it as requiring notice in the fhort hiftory of Conics in his preface.

It may be convenient in this place to make an obfervation on another portion of the preface to Dr. SIMSON's *Conics*, viz. the paragraph beginning, " *Propositionum jamdudum*, &c." (p. vi. Præfat. prim. et fec. edit.) and ending " *contulerint.*"

This paragraph was added in the firft edition, (1735,) in confequence of an objection made by Mr. MAC LAURIN to a Propofition of the Doctor's, as not containing a fufficient acknowledgment of a communication by Mr. M. on the fubject of it. Dr. SIMSON, in a letter to a common friend not named, which remains among his papers, ftates this circumftance. He fays, " I " have not time to write about the Propofition. Mr. MAC LAURIN objects againft. " To take away all controverfy in this matter, I fend you an addition to the " preface;" viz. the above-mentioned paragraph. After fome farther explanation of his anxiety to do juftice to Mr. MAC LAURIN he fays, " you may " fhew him this addition to the preface, before printing it."—The explanation

the nature of this tract, obferving, that it had been erroneoufly attributed to ARCHIMEDES, and it is in an abftract of the works of ARCHIMEDES that he mentions it. He confiders it, however, to be the work of PTOLEMY, without giving any authority. See the above mentioned volume of MAUROLYCUS, fol. 72. It was printed indeed fome years after GOGAVA's publication; but probably MAUROLYCUS had not feen thefe fmall tracts, as he makes no mention of them, or of the editor GOGAVA. The firft tract, *De Parabola*, is fuppofed to be of the time of ROGER BACON, about 1270; but the fecond, *De Speculo Uftorio*, is much more ancient, being quoted by ROGER BACON.

feems to have been quite fatisfactory, for in the following year, 1730, the mathematical correfpondence between Dr. SIMSON and Mr, MAC-LAURIN continued on its ufual friendly footing.

Among Dr. SIMSON's papers there is an account of his *Conics*, drawn up foon after the publication, in the form of a letter, and in the name of fome friend of the Doctor's, apparently with the defign of being tranfmitted to a diftance, and perhaps for publication.* It is a fort of enlargement of the preface, giving a more particular detail of the author's views in compofing the treatife, and of the improvements which he had made in it. The following paffage in it gives a more particular account of the communication to Mr. MAC LAURIN. " The Scholium in p. 198, (firft edit.) and the Prop. it flows from, contain a " very general and ufeful property of the Sections, which the author difcovered " upon this occafion. When Mr. MAC LAURIN, the learned profeffor of " mathematics in the univerfity of Edinburgh, was going to France in the year " 1723, Mr. SIMSON communicated to him a *Locus* of PAPPUS Alexand. " which he had reftored, viz. *Si a tribus punctis datis in recta linea ducan-* " *tur tres rectæ lineæ, et ipsarum intersectiones duæ tangant rectas positione* " *datas, tanget reliqua intersectio rectam lineam positione datam :* which was " afterwards printed in the PhilofophicalTranfactions. When Mr. MACLAURIN " had come home from France, he told Mr. SIMSON that he found, when the " three given points were not in a ftraight line, all the reft remaining as in " the former cafe, that the *Locus* defcribed would be a conic fection, and " that by help of this *Locus* he could eafily defcribe a conic fection through " five given points: This *Locus*, and method of defcription, he gave Mr. " SIMSON without any demonftration, which he faid he had made by help of a " *Lemma* in Sir ISAAC NEWTON's *Principles;* and it was in fearching for " the demonftration of them that Mr. SIMSON found out the Propofitions in " the Scholium, and thofe they flow from, which he did in the fame method

* Since writing this account, Mr. FRYER of Briftol, who is engaged in a hiftory of mathematics, pointed out to me a letter from the celebrated Profeffor HUTCHESON of Glafgow, on Dr. SIMSON's *Conic Sections*, which he had found in the *Bibliotheque Raisonée* for April, May, and June 1735, and which, on comparifon, I afcertained to be the very letter here alluded to. In that journal it is translated into French, and Profeffor HUTCHESON is mentioned as an old friend of M. SMITH, one of the publifhers, at Amfterdam. The letter is dated Glafgow, 28 Feb. 1734.

" they are now printed in his book." The copy of this letter, found among
Dr. Simson's papers, is without any fignature, but it is in the Doctor's
handwriting.*

In the note * page 29, the proper reference is to Boscovich *Sect. Con.
Elementa.* art. 273.

Note B. p. 43.

To the examples mentioned by Dr. Simson, of mifapprehenfions entertained
by eminent mathematicians refpecting Porifms, may be added that of Castillon.
In his commentary on the *Arithmetica Universalis*, (Amft. 1761,) p. 216,
vol. 1, after quoting a paffage from Marinus's preface to the Data, he adds,
" Hæc obfervatio mihi viam aperuit ad reftituenda Euclidis Porifmata;"
and afterwards, page 268, when fpeaking of the ancient problem of the Tactions,
he adds, " omnia foluta habeo, et me editurum fpero propediem cum Porismatis
" Euclidis reftitutis.' The fame author, in his commentary on the appendix,
vol. ii. page 264, fhews more particularly his mifapprehenfion of the nature of
Porifms, by fuppofing them to be the conftructions of Euclid's data; and
then he adds, " ad hunc (fc. librum Porifmatum) non pauca colligere
" cæperam, fed cum audiviffem virum doctiffimum Robertum Simsonum,
" rem perfeciffe, et fua fcripta reliquiffe, ab incepto deftiti, fperans et rogans ut
" tanti viri cogitationes in publicam lucem emittantur." Dr. Simson had
got this book a few years before his death, and was much amufed with thefe
obfervations of the author. It is very furprifing that fo many refpectable
mathematicians deceived themfelves on the fubject of Porifms, efpecially thofe
who were apprized of the failure of Dr. Halley, who candidly acknowledges
it, but who, of all who attempted the inveftigation of them before Dr. Simson,
feemed to be, from his talents and his knowledge, the moft likely to fucceed.

---

* In this letter alfo is given a folution of the problem, " To defcribe a Conic Section
" which fhall pafs through two given points, and touch three ftraight lines given in pofition.'»
—This he eafily deduces from Prop. 12. b. v. (1ft ed.) by means of the before-mentioned
Locus, publifhed in the Phil. Tranfactions for 1723, p. 330, without the tranfformations
employed in Prop. 25 and 26 of Newton's *Princip.* lib. 1.

Note C. p. 22.

" I fubjoin the remainder of Dr. Simson's letter to Dr. Jurin, of which the firſt part is in the text at page 29. " At his (Dr. Halley's) defire I have " confidered that paſſage as narrowly as I could, and ſhall be very glad to have " his opinion concerning the following conjectures; fince, both with reſpect " to the language and the matter, I eſteem him the moſt ſkilful judge.

" The firſt thing to be confidered is, that the figures Pappus ſpeaks of in " this Propofition (for I fuppofe σχῆμαῖος, or fome fuch word, ought to be under- ' ſtood to be joined with ἰσῖα &c.) are fuch as are made by four ſtraight lines, " either all of them interſecting one another, or two of them, at the moſt, " parallels: (the cafe of the parallelogram, becaufe of its plainneſs, being " omitted.) I mean, Pappus ſpeaks of the complex figures made by the " actual interfections of all the lines, and not of fimple Trapezia, whofe fides, " if produced, would meet: this is plain from his fuppofing three points of " interfection in their fides; ' Si dentur, in earum una, puncta tria.'

" Of the three kinds of thefe figures he mentions, that which he calls " παραλλήλον is eafily, and I think certainly, determined, from fome places in " his Lemmas, to be the figure made by two parallel lines meeting two others " which are not; fo that either of thefe two figures A or B is the παραλλήλον.*

" The places which verify this are firſt in Prop. cxxxvi. at the letter C in " page 247† (of the edition at Ven. 1589,) Commandine has it, ' ita KH " ad HL hoc eſt in lineis parallelis GH ad HM.' where the Greek has " τυτ' ἰστιν ἰν παραλλήλω &c. and the figure GHMLK‡ is plainly fuch as has been " defcribed above. Again, in Prop. cxxxv. fol. 246, there occurs, " ut FM " ad DH, ita in lineis parallelis FK ad KH;' which I doubt not will be in " the Greek MSS. ἰν παραλλήλω, and the figure is the fame as in the laſt.

" The fame way of fpeaking is twice ufed in the firſt demonſtration of the " firſt Lemma, fol. 238.§ And all the figures to which thefe two paſſages refer " are of the other kind, (viz. fig. B.) except one. And from this it appears " that παραλλήλον is promifcuouſly applied to both.

* Fig. A and B.   † This ſhould be fol. 246. b.   ‡ Fig. 1.

§ It appears that the MS. ufed by Commandine had ἰν παραλλήλω in this prop. (127.) See his Note D.

"It is alfo plain, that the paffage in fol. 242, at letter B, is not corrupted,
"as COMMANDINE fufpects; It ἰτ δἀὶ γὰρ ἴθκ πρὸς τὴν θι ἰν παραλλήλω, which he
"fhould have rendered into Latin thus, 'ut autem AF ad FG, ita LH ad
'HM, etenim in eadem ratione' (fc. cum utrifque illis rationibus) eft HK ad
'HG ἰν παραλλήλω, viz. eft AF ad FG ut HK ad HG ρ' παραλλήλω AFGHK,"
'and LH ad HM ut HK ad HG, ἰν παραλλήλω LHKGM.'

"Having determined the figure which PAPPUS calls ἀναπαλάω, the reft, viz.
"thofe where none of the lines are parallel, muft belong to other two kinds,
"ὑπτίον and παρύπτικι. The firft of which words ufually fignifies ὑπτιος, re-
"fupinus, reftrorsum vergens, and is applied by ARISTOTLE to the recurved
"horns of Oxen. Now all the figures whatfoever that can be made by the
"interfecting of four ftraight lines none of which are parallel, are with
"refpect to their form, of one kind; and have every one a reverfed or retroflexed
"angle, which will agree very well with the fignification of ὑπτιος reftrorsum
"vergens. But, then, if every one of thefe figures be ὑπτιόι, what comes of
"the remaining branch of the divifion παρύπτιον? This is a difficulty cannot be
"taken away, if we fuppofe PAPPUS to fpeak of the form of thefe figures
"abfolutely confidered; but may be removed, if it be found he has regard
"to the twofold manner in which they are formed, by two lines drawn
"interfecting two other lines, and fo as to give a juft reafon of ufing thefe
"terms ὑπτιον and παρύπτι refpectively.

"I fhall firft explain the two ways after which thefe figures are formed, fo
"as to give rife to the names PAPPUS ufes, and then fhew what ground there
"is to think he confidered them thus, and not abfolutely with refpect to their
"form.

"FIG. 2. Let two lines AB, AC, be drawn, interfecting two others DB, DC;
"then if the angles BAC, BDC, made by thefe lines, lie one over another, i. e.
"ὑπτίοι; or which is the fame thing, if one of them, BDC, be the reverfed
"angle of the figure; then fuch lines may be faid to make the ὑπτιον σχῆμα.

"FIG. 3. But when the two lines AB, AC, are drawn interfecting two
"others DB, DE, fo that the two angles BAC, BDE, made by thefe lines,
"do not lie below one another, but one of them is upon one fide and the
"other upon the other hand of the ὑπτια γωνία AFD, which they form by their
"two neareft fides AC, DE, and fo lie befide (παρά juxta) the reverfed angle,
"as alfo at one another's fides; then thefe pairs of lines may be faid, with
"refpect to the angles they make, to form the παρύπτιον σχῆμα.

" And that Pappus had regard to these two different ways in which one pair
" of lines meet another pair, I think will appear plain enough by comparing
" Prop. cxxxvi. with Prop. cxlii.; in each of which he confiders the figures we
" are now speaking of, as formed by two lines interfecting two others; and thefe
" figures are not only absolutely confidered, exactly the fame; but the things
" demonstrated are also the fame, viz. the converfe of Prop. cxxix.; nor is there
" any thing to give occafion to the fplitting it into two cafes, other than the
" confideration of the different ways the one pair of lines meets the other pair ;
" viz. in Prop. cxxxvi. the lines DH, HE, meet the lines BAE; DAG, ὑπʹλιοι,
" or fo as to form the ὑπʹλοι σχῆμα. And in Prop. cxlii. the lines BD, DE, meet
" the lines AB, AC, παςαλλιοι, or fo as to form the παςαλληλον σχῆμα with refpect
" to their angles.

" Some, perhaps, may incline rather to give παςα the negative or contrary
" force, when joined with ὑπλιοι; becaufe the angles made by the pairs of
" lines which form the figure, are the internal oppofites of two triangles, of
" which the reverfed angle is the common external. I fhall leave this to others
" to determine; and fhall only fay, I do not remember παςα is ufed in this fenfe
" in any mathematical term; for in Euclid's Data I think it is quite amifs
" to tranflate it *contra*."

He adds a Porifm without demonftration, which it is unneceffary to infert,
and concludes, with an apology for writing in hafte,

" I am, Sir, your much obliged and

" affectionate Servant,

" Glasgow, Jan. 10, 1723;   ROBERT SIMSON."

Note [2 C.] p. 49.

On the fubject of Porifms, Dr. Simson was certainly referved; and this
probably in part arofe from a natural defire of publifhing his difcovery of
them in a compleat form, without the rifk of his being anticipated, which
from partial communications was not unreafonably to be apprehended. Can-
dour was a ftriking point of his own character; and we may obferve
in his works the precifion with which he remarks the inventions, the im-

provements, and even the fmalleft hints contributed by his friends;* and perhaps he felt fome diffatisfaction from not having always met with the fame liberality. Some inftances were well known, and there were reports of others; but though they may account for his referve, it is neither of importance to his reputation, nor in any refpect expedient, to inveftigate or to record them His referve on the fubject of Porifms I had frequent opportunities of remarking, though favoured with much familiar converfation with him on mathematical fubjects. From PAPPUS, and other writers, by whom Porifms were mentioned, and alfo from fome very diftant allufions from Dr. SIMSON, I certainly had got a general, but as I afterwards learnt, an imperfect notion of Porifms. I conceived them to be entirely geometrical, and that they were a clafs of Propofitions in which certain points or lines were to be found, which might have a *general* property expreffed in the enunciation. I occafionally fubmitted to Dr. SIMSON fome Propofitions which I confidered to be of that clafs; but without admitting or denying them to be Porifms, with fome pleafantry he faid they were Propofitions. It appears from his *Posthumous Works*, p. 459, that, with his ufual accuracy, a fimple Propofition of that kind (which I had ftated to him as a problem) had been entered in the fair MS. of the Porifms from which the treatife was printed.

NOTE D. p. 52.

As every notice of the Porifms in former times is interefting, I fhall infert fome extracts from PROCLUS refpecting them, in his Commentary on the firft book of EUCLID's *Elements*. I quote from HERVAGIUS's edition, the only printed one, though very erroneous, as is obferved by BAROCIUS in his tranflation, who fays it is " dilaniatum potius quam impreffum." But as BAROCIUS had accefs to feveral MSS. from which he chofe the beft readings, his tranflation, where it differs from the printed Greek, is a better authority; I therefore annex the tranflation, both it and the original being fcarce.

The firft mention of the Porifms is in page 58 of HERVAGIUS's edition, and in page 121 of BAROCIUS, viz. in the Commentary on the firft Prop. of the *Elements*, where he profeffes to give only a brief account of feveral mathe-

---

* See SIMSON's *Conics*, 2d edit. præfat. p. vi. and Prop. 19, lib. 5, at the end. See alfo *Loci Plani*, p. 223.

matical terms, fuch as Lemma, Porifm, Cafe, &c.* The notice of the Porifm is as follows:

" Τὸ δὲ πόρισμα λεγέται μὲν καὶ ἐπὶ προ-
" βλημάτων τινῶν οἷον* τὰ Ἐυκλείδει γεγραμ-
" μένα, σορίσμαλα λεγέλαι δι ιδίως δλαν ἐκ τῶν
" ἀποδεδειγμένων ἀλλο τι συναφανὴ θειώρημα
" μή σροθιμένων ἡμῶν ὁ καὶ δια λιλο πόρισμα
" κεκλήκασι ὥσπερ τι κέρδος ὂν τῆς ἐπιςη-
" μονικῆς ἀποδείξεως σάρεργον."

* δια.

" Corollarium vero, dicitur quidem
" et de quibufdam problematibus ut
" corollaria, quæ EUCLIDI afcripta
" funt. Dicitur autem proprie corol-
" larium, cum ex iis quæ demonftrata
" funt quoddam aliud theorema ap-
" paruerit, nobis minime proponen-
" tibus, quod etiam propterea corol-
" larium vocarunt, tanquam lucrum
" quoddam, quod fit præter gignentis
" fcientiam demonftrationis propo-
" fitum."

This very fhort defcription is unfatisfactory, but a fubfequent paffage (PRO-CLUS, HERVAGII, p. 80. BAROC. p. 173.) is more particular and intelligible:

" Ἐλ τῶν γεωμιτρικῶν ἐςὶν ὀνομάτων το
" πόρισμα. τὖτο δὲ σημαίνει διττὸν. καλῶσι
" γὰρ σορίσμαλα καὶ ὅσα θεωρήμαλα συγκαν
" ἰασπευάζίλαι τᾶις ἄλλων ἀποδείξεσιν διοι
" ἕρμαια καὶ κέρδη τὴν ζηλώλων ὑπάρχοιλα,
" καὶ ὅσα ζηλῖται μὲν ἑυρίσεως δὲ χρήζει
" και ὄυλ γινέσεως μόνης ὄυτε θεωρίας ἀπλῆς.
" ὅτι μὲν γὰρ τῶν ἰσοσκελῶν ἀι πρὸς τῆ βασει
" ἰσαι, θεωρῆσαι.δῖι, καὶ ὄυλ γὰρ δὴ τῶν σραγμαλων
" ἐςιν ἡ τοιάυλη γνῶσις. τὴν δὲ γωνίαν δίχα

" Unum quid geometricorum no-
" minum corollarium eft; hoc autem
" duplex quidpiam fignificat; vocant
" enim corollaria quæcunque etiam
" theoremata, una cum aliorum de-
" monftrationibus probantur, veluti
" lucra inexpectata atque emolu-
" menta quærentium exiftentia: Et
" quæcunque quæruntur quidem, in-
" ventione autem indigent, et neque
" generationis folæ caufa quæruntur
" neque fimplicis contemplationis;
" nam quod quidem ' Æquicrurium
" qui ad bafim funt anguli æquales
" funt,' contemplari oportet, exiften-
" tiumque rerum hujuscemodi cog-

* " Age de iis etiam quæ his annexa funt breviter differamus, quid Sumptio (λῆμμα) " quid Cafus, quid Corollarium (πόρισμα), quid Inftantia, quid Inductio."—BAROC. Verfio, p. 120.

N

NOTES.

"" τιμῖιν ἢ τρίγωνον συσ]ήσασθαι, ἢ ἀφιλῖιν ἤ.
" θίσθαι, ταῦ]α σαι]α σοίησιν τίνος ἀσαιλῆ.
"" τυ δοθέν]ος κύκλυ τὺ κέν]ρον εὑρεῖν ἢ δυο
" δοθέν]ων συμμίτρων μιγαθῶν τὸ μίγισον κ̓αι
"" κοινὸν μίτρον εὑρεῖν ἢ ὅσα τοιάδε μέλα ξύ
"" σως ἱ̓σι σροβλεμάτων κ̓αι θεωρηματ́ων. οὐ]α
" γὰρ γενἱσεις ἱ̓σὶ̓ν ἐν τέτοις τῶν ζηλ̓ωμίνων
" ἀλλ' εὑρίσεις, οὐ]α θεωρία ψ̓ιλὴ. δῖι γὰρ
" ὑσ' ὁ̓ψ̓ιν ἀγαγεῖν, κ̓αὶ σρὸ ὀ̓μμάτων σοιή-
" σασθαι τό ζηλ̓όμενον. τοιαῦ]α ἄρα ἱ̓σι κ̓αὶ
" ὅσα σορίσμαῖα Εὐκλείδης γίγραφ̓ι, βιβλία
" σροβλημαῖων* συν̓ίαξας. ἀλλά σερὶ μὲν τῶν
" τοιούτων σορισμάτων σαρίισ]ω λεγειν. τὰ δὲ
"" ἐν τῇ σοιχειώσι σορίσμαῖα συναναφαίνι]αι
" μὲν τἀῖς ἄλλων ἀσοδεί ξισιν, αὐ]α δὲ σρον-
" γυμίνης τυγχάνι ζηλή̓σεως. δ̓ιον κ̓αι τὸ νὺν
" σροκ̓ιμ̓ινον." &c.

" nitio eſt. ' Angulum autem bifa-
" riam ſecare,' vel ' triangulum con-
" ſtituere,' vel ' rectam lineam æqua-
" lem abſcindere, vel ponere,' hæc
" omnia ut aliquid fiat poſtulant.
" ' Dati vero circuli centrum reperire,'
" vel ' duabus magnitudinibus com-
" menſurabilibus datis, maximam ip-
" ſarum communem menſuram in-
" venire,' vel quæcunque id genus
" alia, quodammodo inter proble-
" mata atque theoremata ſunt; neque
" enim quæſitorum ortus in his,
" neque ſola contemplatio ſed inventio
" eſt, Opus eſt ſiquidem quæſitum
" in conſpectu et præ oculis ponere.
" Talia igitur ſunt quæcunque etiam
" corollaria EUCLIDES ſcripſit, quippe
" qui libros corollariorum conſtruxit.
" Verum de hujuscemodi quidem co-
" rollariis dicere prætermittatur. Quæ
" autem in elementari inſtitutione ſunt
" corollaria, ſimul quidem cum alio-
" rum demonſtrationibus apparent,
" ipſa vero non præcipue quæruntur,
" veluti id quod in præsentia propo-
" nitur ;" viz. Cor. 15. 1 Elem.

† From the ſenſe, this word ought to be σορίσμαῖων, and not σροβλήμαῖων; for whatever be the origin of that claſs of Propoſitions, there can be no doubt of their being always known and diſtinguiſhed by the term σορίσμαῖα. BAROCIUS accordingly tranſlates it in this place by " corollariorum;" moſt probably from his having found that reading in a MS. which he relied upon. In the MS. of PROCLUS, in the Bodleian Library, which is eſteemed valuable, σροβλήμαῖων is in the text, but on the margin σορίσμαῖων is written in a character much reſembling the text. In this ſentence γ probably is omitted after βιβλία.

NOTES. 91

In this extract is a very explicit statement of the two very different species of *Porisms*, (or *Corollaries*, as they are named by BAROCIUS,) viz. the Porisms composed by EUCLID, a curious and difficult class of Propositions, requiring investigation, as well as construction and demonstration; and the Porisms or corollaries of EUCLID's Elements, which result from the demonstration of other Propositions, and which often present themselves unexpectedly. PROCLUS proceeds to illustrate this latter class of Porisms by the example of a corollary annexed to 15 Prop. 1 Elem. the preceding extract being part of his commentary on that Proposition. He then gives a more diffuse description of this corollary of the elements, of which a portion is annexed.

(HERVAGII. p. 80.

" "Εςιν δὲ τὸ πόρισμα θεώρημα διὰ τᾶ ἄλλω
" προβλήμαίος ἢ θεωρήμαίος ἀποδείξιως,
" ἀπραγματεύίως ἀναφαινόμενον. δίρη γὰρ
" κἀλα τύχην πυρωπίπἰειν ἐοίκαμεν τοῖς πο-
" ρίσμασιν, ὃ γὰρ προὔμμίνοις, ἀδὲ ζηπίσπὀνη
" ἀπαπλῖ, ὅθεν αὐτὰ καὶ τοῖς, ἱρμάιας
" εἰκάσαμεν, καὶ ἴσως οἱ δεινοὶ τὰ μαθημα-
" ἴκὰ, καλα τάυτην αὐποῖς ἔθεπο την ἐπωνυμίαν,
" ἐνδεικνύμενοι τοῖς πολλοῖς καὶ ἐπὶ τὸ φαι-
" νόμενον κέρδος ἐπλοημμίνοις* ὅτι ἄρα τὰ
" ἀληθῆ θεῦ δῶρα καὶ λα ἱρμαια τάυλα ἰςίν
" ὐχ οἶα ἐκείνοις δοκῖ."

* Forsan ἠδομίνοις.

BAROC. p. 174.)

" Corollarium est theorema quod
" ex alius problematis vel theorematis
" demonstratione ex improviso emer-
" git: nam veluti casu quodam in
" corollaria incidere videmur, nec
" proponentibus enim nobis, neque
" etiam quærentibus obviam sese
" offerunt. Unde hæc quoque lucris
" affimilavimus: et fortasse mathe-
" maticarum rerum periti hoc ipsis
" imposuere nomen, ostendentes vul-
" go, quippe quod apparenti gaudet
" lucro, quod utique vera DEI mu-
" nera, veraque lucra hæc sunt, non
" autem quæ illi videntur," &c.

Then follow several distinctions of corollaries, into arithmetical and geometrical, into those arising from theorems and those from problems, with some others of no particular importance. These extracts, it must be allowed, are not expressed with uniform clearness, but they discriminate sufficiently the Porisms of EUCLID from the corollaries of the Elements; and they correspond with the more general expressions of PAPPUS on this subject, which PROCLUS in these passages plainly had in view. For illustrating the

N 2

the diftinction, he premifes that the 5th Prop. 1 Elem. is a theorem, and requires demonftration : alfo, that the 1ft, 3d, and 9th Propofitions are problems, each requiring fomething to be done or conftructed; he then adds, that to find the centre of a given circle, (1. 3. Elem.) or to find the greateft common meafure of two magnitudes, (2. 7. Elem.) or fuch like things, are in fome fort of an intermediate character between problems and theorems. For the conftruction of thefe things fought is not given in the enunciation, but invention is requifite for finding that conftruction, and alfo for difcovering the demonftration; for it is neceffary to exhibit to the eye the conftruction of the things fought. Such, fays he, is the nature of the fecond kind. of Porifms, of which EUCLID compofed books.* Of thefe obfervations, not altogether clear, I have given, in a fort of paraphrafe, the meaning as I underftand it to be. It is plain he was acquainted with the obfervation of PAPPUS, that Porifms are of a middle nature between problems and theorems, though fome doubt may be entertained, whether the examples which he quotes be properly ftated as Porifms, as was obferved in a former note p. 47. It may be remarked here, that PROCLUS, in referring to the 2d. 7th Elem. as a Porifm, fuppofes that there may be arithmetical Porifms as well as geometrical, and this Propofition is altogether unconnected with *Loci*, or even with geometrical pofition of any kind.

I may obferve further in this place, that *Loci*, which by PAPPUS are reckoned a clafs of Porifms, have alfo fomewhat of the intermediate character between problems and theorems; though they are generally reckoned to belong to the latter clafs. In the *Locus* the conftruction muft be inveftigated, as in the *Porism*; and every *Locus* is eafily convertible into a Porifm.

The etymology of the Porifm, or corollary of the Elements, from πόρισμα fignifying gain, may be right ; and the other meaning of the Greek term πόρισμα, implying inveftigation, may be the ground of its being applied to the Porifms of EUCLID. In the firft fenfe it is the common corollary, which is an acquifition (or gain) from another Propofition, from the demonftration of which it refults often unexpectedly.† Πόρισμα alfo, from the other fignification of the word, properly denotes

* See Note, p. 90.

† Corollarium, affumed by BAROCIUS as the proper tranflation of πόρισμα, may exprefs the firft meaning of it, but has no connection with the other.

I

any thing to be inveſtigated, which correſponds with the character of the Poriſms of Euclid.* And thus, without any connection between theſe two claſſes of Propoſitions, they may incidentally, from the two unconnected meanings of a Greek word, have obtained the ſame name.

It is neceſſary, however, to remark, that the firſt extract from Proclus, (p. 58, Hervag.) containing the very ſhort deſcription of Poriſms, does not, as it ſtands, accurately correſpond with the detailed and more intelligible deſcription in the ſecond. If the word προβλημαίων in the former paſſage be the true reading, it might thence certainly be urged, that Poriſms have a reference to problems; and an argument might be derived in favour of the opinion, that the ancient notion of Poriſms was ſimilar to that propoſed by Mr. Playfair. But this preliminary deſcription of Poriſms by Proclus is ſurely too ſhort and general to be the foundation of any ſuch inference, without other more preciſe authorities; and the ſubſequent paſſage, which is much more particular and intelligible, contains no ſuch reference to problems. Beſides, it is well known, that the only printed edition of Proclus is very erroneous.† If it may be ſuppoſed that προβλημαίων is printed for πορισμαίων,‡ as it muſt certainly is in another place in the ſecond extract, the two paſſages will be more conſiſtent; and though the expreſſion in the firſt be ſhort and inexplicit, it

---

* Notwithſtanding theſe notices of Euclid's Poriſms, Proclus does not ſpeak of their curioſity and importance in the manner that Pappus does. And it is ſomewhat remarkable, that when Proclus gives a catalogue of the works of Euclid, (p. 20, Hervag. edit.) he makes no mention whatever of his Poriſms. Dr. Simson, moſt probably, had conſidered theſe paſſages in his firſt enquiries on this ſubject; but as they could not give the ſmalleſt aid towards his diſcovery of the true nature of Poriſms, he does not ſeem afterwards to have attended to the remarks of Proclus, as there is no alluſion to them among his papers.

† Barocius, in the dedication of his tranſlation of Proclus, ſays, " quamvis neſcius non " eſſem quod impreſſi fuerant Basiliæ quatuor Procli Diadochi libri commentariorum in " primum Elementorum Euclidis ; quos adeo laceros et corruptos vidi, ut nihil boni ex iis " elicere potuerim : editi namque erant perinde ac ſi editi nunquam fuiſſent." Afterwards, having mentioned the different MSS. to which he had acceſs, he adds, " ubi ex iis omnibus " exemplaribus quoad fieri potuit unum integrum féci quod poſtremo e Græca lingua in " Latinam converti, &c."——Proclus Barocii, Patav. 1569.

‡ I muſt obſerve, that in the tranſlation by Barocius, which, though it includes many corrections of the printed Greek, yet contains many errors, the MS. which he relied on, muſt have had προβλημαίων in this place, as in the edition by Hervagius.

being only an introductory notice of Porifms, yet it may be underftood. I conceive, therefore, that in both extracts his object is the fame, viz. to diftinguifh the Porifms compofed by EUCLID from the Porifm which is the common corollary; and he does it on the fame principle, though more briefly in the firft than in the fecond: and if σορίσμῖων be affumed as the true reading in the firft paffage, it may be freely tranflated thus: " The term Porifm is applied to " that clafs of Propofitions of which are the Porifms compofed by EUCLID; " but more properly the term Porifm is applied to Propofitions not propofed by " us, but arifing from the demonftration of other Propofitions; and being thus " an unexpected acquifition- they obtained that name."

It may be further obferved, that in the Commentary of PROCLUS on the 1ft, 1. Elem. there are various diftinctions of problems ftated; one of which, viz. the πλεονάζον and the ἔλλειψις may here be properly mentioned. By the former term (exceeding) is meant any problem which has more conditions or data than are neceffary to the folution. As for inftance, if it be propofed to defcribe an equilateral triangle, of which each of the angles fhall be two thirds of a right angle. This laft condition is implied in the former, and is therefore fuperfluous. And if an unneceffary condition be added, the probability is that it will be inconfiftent with the others, and render the problem fo ftated impoffible. The other term (deficient) expreffes the want of a neceffary condition, by which the problem becomes, according to modern language, indeterminate.* Ἐλλείψεις δὲ

---

* In PAPPUS and PROCLUS the term ἀδιόριςος, commonly tranflated indeterminatum, or indeterminate, is applied to problems which do not require or admit of any determination, a meaning altogether different from the modern ufe of the word: ἀορίςος is indeed ufed by PROCLUS in the modern fenfe of indeterminate. The following fentence may be quoted, HERVAGIUS, p. 61, and BAROC. 127, at the clofe of the commentary on the firft propofition of the Elements: " φανερὸν οὖν ἐκ τούτων ὅτι τὰ κυρίως λεγόμενα προβλήματα, βούλεται " τὴν ἀορισίαν διαφεύγειν, καὶ μὴ εἶναι τῶν, ἀπειραχῶς γιγνομένων. λέγεται δὲ ὅλως κἀκεῖνα προ- " βλήματα διὰ τὴν ὁμωνομίαν τε προβλήματος." &c. " Ex his itaque manifeftum eft, quod ea " quæ proprie problemata appellantur, indeterminationem effugere debent, et non effe ex " eorum numero quæ infinitis modis fiunt. Problemata tamen et illa dicuntur per problematis " æquivocationem, &c." BAROC.

In this paffage it is intimated that indeterminate Propofitions are alfo called problems, from a double meaning of the word; and among fuch problems Loci muft of courfe have a place, and therefore Porifms alfo; but it is not eafy to reconcile the different obfervations of PROCLUS on this fubject.

ἔστι πρόβλημα, τὸ προσθήκης ἄλλης δεόμενον ἵνα ἐκ τῆς ἀορισίας, εἰς τάξιν καὶ ὅρον ἐπιστημονικὸν ἄχθη.† HERVAG. edit. p. 61.

If any fuch notion had exifted in the times of PAPPUS or PROCLUS, as the ingenious definition of Mr. PLAYFAIR involves, we might have expected fome reference or allufion to it in a difcuffion like this concerning problems. PROCLUS, in the prefent ftate of the text, even with all the fkill of BAROCIUS, is not always intelligible, nor confiftent in his different obfervations on the fame fubject. See his comment on the 1ft Prop. EUCL.; alfo on the 22d Prop. compared with his comment on the corollary to the 15th Prop.

NOTE E. p. 57.

The 117th Prop. of the feventh book of PAPPUS has become remarkable, from the confideration which has been given to it by many diftinguifhed. mathematicians of modern times, more than from any particular excellence belonging to it, as it ftands in that author. In 1742, an extenfion of it was propofed by M. CRAMER of Geneva to M. DE CASTILLON, viz. by fuppofing the three points not to be in a ftraight line, as is affumed in the Propofition of PAPPUS; and then the problem was thus expreffed, "a circle and three " points being given in pofition, to infcribe a triangle in the circle of which " the three fides fhall refpectively pafs through the three given points." It was not however till 1776, that CASTILLON publifhed a folution, in the Berlin Memoirs of that year. The folution is geometrical, and diftinguifhed into many cafes. In the fame volume is an algebraical folution by LA GRANGE, by means of fome trigonometrical formulas. In the *Petersburgh Acts* for 1780, are folutions of the fame problem by Meffrs. EULER, LEXEL, and FUSS. In the *Memorie di Mathematica e Fisica della Società Italiana*, tom. iv. Verona, 1798, are two memoirs refpecting this problem.* The firft is by GIOR-

---

† HUGO DE OMERIQUE, in the fourth book of his *Analyſis Geometrica*, feems to have had fome of the diftinctions of PROCLUS in view. The *Deficiens Problema* of the latter is called *diminutum* by the former, who gives fome examples of fuch problems, in which the folution turns out to be a *Locus*.

* The following folutions and extenfions of this problem were pointed out to me by Mr. PLAYFAIR.

DAMO DI OTTAJANO, in which is a sketch of the history of the problem; and though his solutions are geometrical, yet he appears to have had but an imperfect notion of the ancient geometrical analysis. He resolves the problem before mentioned, adding a variation of it, viz. that the three lines from the angular points of the inscribed triangle to the three given points may make angles with the sides of the triangle equal to given angles; and concludes with an important extension to the case of a polygon, of any number of sides, which he inscribes in a given circle, so that the sides respectively shall pass through the same number of given points. The other memoir in the same volume is by Signor GIAN. FRANCESCO MALFATTI, and contains a good solution of the very general problem last mentioned; from reducing the case of polygons to that of the triangle, by the use of two very easy theorems. I should have added to the account of the first memoir, that the author acknowledges his inability to resolve the problem algebraically, and expresses a wish that some mathematician would attempt a solution purely analytical. Hence also this problem may be considered as an example of what Dr. SIMSON observes, that many geometrical problems are resolved more easily by the ancient analysis than by the modern.

In the Berlin Memoirs for 1796, is a memoir by LHUILIER, containing an algebraical solution of the most general case of the polygon. He premises two trigonometrical theorems, one of which is the basis of La GRANGE's solution of the case of the triangle, and by means of them proceeds successively from the simple to the more complex cases of polygons; and in all the cases deduces a quadratic equation, from the solution of which the problem is resolved. The expressions of course become very complex; and a geometrical construction derived from them would be not less so. LHUILIER extends the problem also to Conic Sections, and adds a similar one respecting the sphere.

I shall subjoin Dr. SIMSON's solution of the case, when the three points are not in a straight line; dated August 30, 1731, many years before this problem acquired celebrity from having employed the skill of so many eminent mathematicians. The Doctor wrote all his mathematical notes in Latin, and I give this Proposition in his own words.*

† In this and some other Propositions of Dr. SIMSON's copied into these notes, the Data are quoted as numbered in the old edition. I may remark also, that there are some differences in the phraseology of Dr. SIMSON's early writings from what he used in more advanced life, but they are not important.

("Eſt Prop. 117, lib. vii. generalior facta.") Fig. 4.

" Datis tribus punctis A, B, C, et circulo DEF poſitione dato,
" a duobis ex punctis A, B, inflectere ad circumferentiam AE, EB
" occurrentes circulo in D, G, quæ faciant DGC, rectam lineam."

" Factum puta ; et ducatur DF parallela ipſi AB junctæ, occurrens circulo
" in F, et juncta FG occurrat AB in H. Quoniam igitur propter parallelas
" eſt angulus BHF æqualis angulo DFG, hoc eſt, propter circulum, angulo
" DEG, erunt triangula AEB, BHG æquiangula, et igitur rectangulum ABH
" æquale erit dato rectangulo (92 Dat.) EBG, et datur AB, quare et BH et
" punctum H, et igitur juncta HC poſitione dabitur ; occurrat hæc ipſi DF
" in K, et erit angulus DKH æqualis dato angulo BHK ; quare eo deventum
" eſt, ut a duobus punctis datis C, H, inflectantur ad circumferentiam CG GH,
" ita ut DF datum faciat angulum DKH cum ipſa HC ; puta factum, et
" ducatur DL parallela ipſi CH, et juncta LF occurrat CH in M, ergo dabitur
" punctum M, ut prius oſtenſum de puncto H ; [ſc. quoniam angulus CMF
" æqualis eſt angulo FLD, hoc eſt angulo FGC, igitur æquales ſunt anguli
" HMF HGC, et triangula HMF, HGC, ſimilia, quare et CHM rectangulum
" dato FHG eſt æquale, et datur GH, ergo et MH et punctum M,] et propter
" datum angulum DKH dabitur angulus KDL, et igitur recta LF magnitudine
" dabitur [88 Dat.[quoniam igitur a dato puncto M ad circulum poſitione
" datum ducta eſt MFL, faciens FL quæ circulo interœpitur datam, dabitur
" ML poſitione, et igitur LD et AD, et BE. Q. E. I.
" Compoſitio.

" Fiat rectangulum ABH æquale rectangulo contento ſegmentis cujuſvis
" rectæ quæ a puncto B ad circulum producitur, et jungatur HC, fiat vero
" rectangutum CHM æquale rectangulo contento ſegmentis cujuſvis rectæ
" quæ a puncto H ad circulum producitur, et a puncto M ducatur MFL quæ
" abſcindat ſegmentum LEDF capiens angulum angulo CHB æqualem, junctaque
" HF occurrat circumferentiæ in G, juncta vero BG eidem occurrat in E, et
" juncta AE eidem occurrat in D. Erunt puncta D, G, C, in recta linea. Jungatur
" DF, et connectentur DL, DG, CG : et quoniam rectangulum ABH æquale
" eſt ex conſtructione, rectangulo EBG, erit angulus BHG æqualis angulo
" E hoc eſt angulo DFG, igitur parallelæ ſunt AB, DF, ergo angulus DKH

o

" æqualis eft ipfi BHK, hoc eft ex conftructione angulo LDF quare parallelæ
" funt DL HK; et propter rectangulum CHM æquale rectangulo FHG in
" circulo funt puncta C,M,G,F; ergo angulus CGF æqualis eft [angulo
" CMF, hoc eft] ipfi DLF, et propter circulum anguli DLF et DGF
" fimul æquales funt duobus rectis, ergo anguli CGF, DGF, fimul æquales funt
" duobus rectis, et igitur recta eft linea DGC. Q. E. D."—" Tranfcriptum
" Aug. 30, 1731."

It may alfo be mentioned that Dr. SIMSON at the fame time gave two
variations of the problem, viz. the fame things being given to make the line
DG parallel to a given ftraight line; and alfo the fame things being fuppofed,
to inflect AEB, fo that the line DG may make an angle with the line drawn
from D to the third given point, equal to a given angle.

NOTE F. p. 58.

The cafe of inclinations refpecting the Rhombus has always been confidered
as an elegant geometrical Problem, and it attracted Dr. SIMSON's attention
in his early ftudy of PAPPUS. His amendment of the figure of a Lemma
in COMMANDINE's PAPPUS, (Prop. 70. lib. vii. PAPPI,) which no doubt
was ufed by APOLLONIUS in the folution of this problem, and the Doctor's
correfponding folution of it, are dated Feb. 20, 1723. The following
tranfcript was taken from a later copy without date, but is fubftantially the
fame with the original paper, with fome flight amendments.

It is proper, however, to mention, that the Problem of the trifection of an arch
of a circle is treated of in the iv. b. of PAPPUS ; and in Prop. 31ft it is propofed to be done by an inclination which is a folid Problem, there refolved
by an hyperbola: The cafe of inclinations employed is that of a rectangular
parallelogram; though the folution be the fame in any other parallelogram. Of
courfe, a Rhombus being a parallelogram, it is alfo comprehended in the
general folution. But this cafe of the Rhombus is a Problem in the treatife
by APOLLONIUS De Inclinationibus, and is plane. Dr. SIMSON,
however, in his notes on this Prop. of PAPPUS fhews, that when the Hyperbola
is employed in the cafe of the Rhombus, the requifite circle is defcribed
about the vertex of the Hyperbola, and then the interfections of the Hyperbola

aiid circle may be determined by plane geometry, and without the defcription
of the Hyperbola. Thus the Problem of the Rhombus, though a cafe of a
folid Problem, as often happens, becomes plane; and is refolved in this manner
by Dr. SIMSON. But the following folution of the Doctor's is better, being
no doubt the APOLLONIAN, as the lemma, Prop. 70. vii. PAPPI, is employed.

"*Prop. de Rhombo cui inservit Prop: 70. lib. vii.* PAPPI."

[" Schemate hujus fc. Prop. 70, emendato ut in Fig 6."].

" Rhombo existente, ABCD, et producta AC ad E, facere EF
" datam, et quæ ad punctum B vergat. Fig. 5."

" Factum puta et producta BC ad G, angulo FCG æqualis fiat BFG, et
" GE jungatur. Ergo angulus GFE, æqualis eft ipfi [FCB hoc eft, ipfi BCA
" feu] GCE. Sunt igitur F, C, G, E, in circulo; quare angulus FGE,
" æqualis eft [ipfi FCE vel ei quæ ipfi deinceps eft, hoc eft] angulo CDB :
" et in triangulis FGE, BDC, oftenfus eft GFE æqualis angulo DCB
" æquiangula propterea funt triangula. Ergo ut BC ad BD, ita EF ad
" EG, et data ex hypothefi eft FE, quare data eft EG, five ipfi æqualis
" FG. Eft autem propter æquiangula triangula BGF, FGC, BG ad GF
" ut GF ad GC; igitur rectangulum BGC æquale eft quadrato ex GF :
" datum proinde eft BGC rectangulum, et data eft BG ergo [84 Dat.].
" dabitur CG, et punctum G propterea datum erit; et data eft GF magni-
" tudine, et CD pofitione ergo (31 Dat.) pofitione data eft GF et punctum
" F, fed et punctum B, ergo recta BFE pofitione data eft.

" Componetur ita : Sit BH data recta, et ut BC ad BD, ita fiat BH ad
" quartam BK : ad datua vero rectam BC applicetur rectangulum BGC
" æquale quadrato ex BK, excedens quadrato ; factaque GL ipfi BK
" æquali; cetitro G intervallo GL defcribatur circulus qui rectis AC, CD
" occurrat in E, F, punctis : Erit juncta EF æqualis datæ BH vergetque ad
" punctum B.

" Jungantur enim G E et G F; et quoniam GL media eft proportionalis
" inter BG, GC, erunt puncta E, F, B, in recta linea per Prop. 70. lib. vii.
" PAPPI. Et quoniam eft BG ad GL feu GF ut GF ad GC, æquiangula erunt
" triangula BGF, FGC; quare angulus BFG æqualis eft ipfi FCG; et angulus

" GFE ipfi [FCB hoc eft, ipfi BCA feu] GCE. Ergo in circulo funt punɑ̂a F,
" C, G, E, angulufque FGE æqualis erit [ipfi FCE vel ei qui ipfi deinceps eft
" hoc eft] angulo CDB : æquiangula igitur funt CBD, EFG triangula, et ut
" CB ad BD, hoc eft ut BH ad BK, ita erit EF ad EG vel FG ; æqualis autem
" eft BK ipfi FG feu GL, ergo æqualis eft BH ipfi EF. Q. E D.

" Potuiffet compofitio æque ac analyfis faɑ̂a fuiffe fine PAPPI Lem-
" mate, fc. jungendo BF et producendo eam ad AC, in E; ita enim non opus
" fuiffet Lemmate ad oftendendum punɑ̂a E, F, B, in reɑ̂a linea effe. Hæc
" autem compofitio ftriɑ̂e loquendo refpondiffet problemati quo ,requiritur a
" punɑ̂o B ducere reɑ̂am BFE, et facere EF datam ; et non huic quo re-
" quiritur producere AC ad E, et facere EF datam quæ ad punɑ̂um B vergit ;
" hoc eft in angulo exteriore Rhombi ECF, feu interiore ACF, aptare reɑ̂am
" EF magnitudine datam quæ vergat ad punɑ̂um B ; et ut compofitio huic
" enunciationi ad amuffim refpondeat videtur Lemma PAPPI utile, et non
" aliam ob caufam. R. S."

" In cafu anguli interioris Rhombi, addatur determinatio ope, Prop. 73, 74,
" lib. vii. PAPPI, et fiant figuræ refpondentes pro cafu, [Prop. 70. PAPPI,
" et inde in problemate] quando ponenda EF in angulo interiore ACD, et
" etiam per B punɑ̂um ; [vel ad B vergens.]

" Lemma in linca ultima, fol. 205, a. PAPPI, (COMMAND. 1589) assump-
" tum ab eo, eft ut sequitur. Vid. fig. Prop. 70. ' Si a punɑ̂o C in diametro.
' DF ducantur ad circumferentiam reɑ̂æ CL CK, quæ faciant cum diametro.
' æquales angulos LCF, KCF, erunt duɑ̂æ inter fe æquales ;' cujus demon-
" ftratio facilis eft. Vide fig. 6.[Schema emendatum, Prop. 70. lib. vii. PAPPI."]

NOTE G. p. 62.

As Dr. SIMSON had at one time intended to compofe a treatife on *Loci
ad superficiem*, it may be fatisfaɑ̂ory to fome readers to fee his preface, in
which he expreffes this intention ; and this, with fome excerpts from the
remaining fragments on this fubjeɑ̂, may give a general notion of the Doɑ̂or's
views of this fubjeɑ̂.

" De *Locis ad superficiem*, quæ ab EUCLIDE aliifve geometris antiquis
" tradita fuere, injuria temporis deperdita funt, exceptis paucis iis quæ apud

" Pappum habentur. Recentiores* autem nihil de *Locis* hifce cognofcere
" videntur, exceptis *Locis* quæ funt ad fuperficiem rotatione alicujus ex fec-
" tionibus conicis circa diametros vel alias rectas genitam, cum tamen
" magnus fit numerus aliorum qui hoc modo minime formari poffunt. Operæ
" igitur pretium facturum ab eruditis videri fpero, fi viam qua *Loci* hi invefti-
" gari poffunt exemplis quibufdam patefaciam, quibus rite intellectis, non
" difficile erit ei qui doctrina *Locorum solidorum* probe inftructus fuerit,
" *Locos* ad quamvis fuperficiem, quæ quovis modo ex fectionibus conicis
" vario fitu difpofitis oriri poteft, inveftigare et defcribere.

" Hi autem omnes [phrafi algebraica] fequente æquatione continentur, viz.

$$ \text{" } xx + \frac{a}{b}xy + ax + by + \frac{a}{c}yy + \frac{a}{d}yz + \frac{a}{e}xz + gz + \frac{b}{k}zz = \text{ Dato.} $$

" Signis utcunque mutatis."

The firft example, as was obferved in the Memoir, is unfinifhed. The
Doctor had proceeded fo far as to afcertain the common fections of the fuperficies required with feveral planes; fome others were propofed, but not
inveftigated, and there is no trace of his having refumed the inquiry. The
portion of this example which is written becomes not a little complex, requiring alfo fubfidiary Propofitions; and therefore the enunciation only of the
Locus is here ftated; which may be referred to the diagram of the fubfequent
Propofition. [Fig. 8.]

" Data sint in plano quovis, puta in plano chartæ in quo
" schema est, duo puncta A, B, et juncta BA, in ea sumatur
" quodvis punctum D, et ipsi AD ad rectos angulos, ducatur in
" eodem plano recta DE, et a termino ejus E ad planum erigatur
" ad rectos angulos EF, sitque rectangulum BDA simul cum
" rectangulo DEF, et quadrato quod fit ex EF æquale spatio dato,
" puta (simplicitatis gratia) quadrato ex ipsa AB : quæritur
" superficies in qua omnia puncta F versantur."

To this no equation is annexed.

* " N. B. fcripta fuerunt hæc priusquam librum, D. N. *des Courbes a double Courbure*
" uidiffem, itaque mutanda eft hæc introductio."

The following is a short example of a *Locus ad superficiem*, found among several unfinished sketches of such Propositions which were mentioned in the Memoir. It is given as it stands in the Doctor's own words.

" *Locus ad Superficiem.* $xx + \frac{ay^2}{b} + \frac{az^2}{c} = d^2$ [FIG. 8.]

" Data positione recta linea AB, punctoque in ipsa A, in
" dato positione plano ABC; si a puncto D in recta AB ipsi
" ad rectos angulos ducatur recta DE, et a termino ipsius E,
" erigatur ad planum ABC perpendicularis recta EF ; fueritque
" quadratum ex AD, simul cum spatio quod ad quadratum
" ex DE, et eo quod ad quadratum ex EF, datas habent rationes,
" æquale dato spatio, scil. quadrato ex data recta AB.   Tanget
" punctum F, superficiem positione datam."

" Sit enim quadratum ex DG, id quod ad quadratum ex DE datam
" habet rationem, et juncta AG occurrat circulo centro A. femidiametro AB
" descripto in H, idem vero circulus occurrat DE in K, L, punctis.   Quoniam
" igitur quadrata ex AD, DG, hoc est quadratum ex AG, una cum eo quod
" ad quadratum ex EF datam habet rationem æquale est quadrato ex AB, seu
" AH, auferatur commune quadratum ex AG, et reliquum rectangulum KGL
" æquali erit spatio quod ad quadratum ex EF datam habet rationem.   Fiat
" autem ut DG ad DE, ita DK ad DM; quadrata igitur ex hisce rectis pro-
" portionalia erunt, et [per. 19. 5.] quoniam quadratum ex DK est ad quadratum
" ex DM ut quadratum ex DG ad quadratum ex DE, erit in eadem ratione
" data reliquum rectangulum KGL ad excessum quadratorum ex DM, DE,
" hoc est, facta DN æquali ipsi DM, ad rectangulum MEN.   Datur igitur
" ratio rectanguli MEN ad rectangulum KGL; hujus autem ratio ad qua-
" dratum ex EF datur, quare [per 8. Dat.] datur ratio MEN ad quadratum ex
" EF.   Quoniam vero datur ratio DK ad DM, tangit autem punctum K
" circumferentiam positione datam, femidiametro, scil. AB descriptam, tanget
" punctum M ellipsin positione datam, cujus femiaxis major est ipsa AB;
" occurrat vero EF sphæroidi oblongæ hac ellipsi descriptæ in O.   Et quoniam
" data ostensa fuit ratio rectanguli MEN, hoc est quadrati ex EO ad quadratum
" ex EF, dabitur ratio EO ad EF.   Si igitur altitudo EO omnium punctorum
" O hujus sphæroidis supra planum ABC diminuatur, vel augeatur, in ratione

" data EO ad EF fuperficies folidi, diminutione hac vel additione producti erit
" Locus punctorum F. Q. E. I."

Another fhort Propofition exhibits a method of defcribing the Caffinian
curve by means of *Loci ad superficiem.*

PROB. [Fig. 9.]

" Describere curvam Cassinianam."

" Sint A, B, Foci, C centrum, et D junctum in curva; igitur junctis AD
" BD, erit rectangulum ADB æquale dato fpatio."

" Plano in quo eft curva ducatur ad rectos angulos recta DE ipfi DA
" æqualis, et tanget punctum E conum rectangulum cujus vertex eft punctum
" A, axis vero plano ABD ad rectos angulos ; et quoniam rectus eft angulus
" EDB, et rectangulum EDB æquale dato fpatio, tanget idem punctum E
" hyperbolam æquilateram cujus una affymptotos eft DB, altera vero recta
' BF quæ ad planum ADB eft perpendicularis. Revolvatur igitur hyperbola
" hæc circa BF tanquam axem, et defcribetur folidum hyperbolicum acutum ;
" communis vero fectio fuperficiei hujus cum fuperficie coni prædicti, erit linea a
" cujus punctis fi ducantur perpendiculares ad planum ADB, ipfi occurrent
" in punctis quæ funt in curva Caffiniana, ut patet."

Another example may be the 28 Prop. lib. iv. PAPPI, as corrected by Dr.
SIMSON.

PROB. v. PROP. 28. Lib. iv. PAPPI. [Fig. 7.]

" Hic igitur lineæ ortus magis mechanicus est, ut dictum
" fuit, geometrice vero per Locos qui ad superficies " dicuntur"*
" resolvi potest hoc modo."

" Sit circuli quadrans pofitione datus ABC, et ducatur, ut contingit, recta
" linea BD, et ad BC perpendicularis EF quæ ad circumferentiam DC pro-
" portionem datam habeat. Dico punctum E ad lineam effe.

* Deeft hoc verbum in MS. BULL.

"Intelligatur enim a circumferentia ADC recti cylindri fuperficies, et
"in ipfa linea fpiralis [CGH] defcripta, [incipiens a puncto C, ita ut
"velocitas puncti defcribentis in recta CM quæ perpendicularis eft ad
"planum ACB, fit ad velocitatem ipfius CM in quadrante circuli CDA in
"data ratione quam habet EF ad circumferentiam DC, quæ propterea
"fpiralis] pofitione data erit. Sitque HD latus cylindri, et ad planum
"circuli perpendiculares ducantur EI, BL, et per H ipfi BD parallela
"ducatur HL, [ipfi EI occurrens in I.] Itaque quoniam proportio rectæ lineæ
"EF ad circumferentiam DC eadem eft quæ proportio EI, hoc eft DH
"ad eandem DC propter fpiralem; erit recta EF ipfi EI æqualis. Suntque
"FE, EI, pofitione, ergo et juncta FI pofitione erit, [N. B. Non dicit FE, EI,
"pofitione datas, datum non eft punctum E, fed intelligit tantum angulum
"EFB datum effe, et EI perpendicularem effe ad planum circuli, unde
"quoniam rectus eft angulus EFB, erit :] et FI ad BC perpendicularis. Eft
"igitur FI in plano [fc. quod per BC tranfit, et per dimidium anguli recti
"inclinatum eft ad planum fubjectum ABC, angulus enim EFI dimidium
"eft recti] quare et punctum I eft in eodem; atque idem punctum eft in
"fuperficie, fertur enim HL et per lineam fpiralem CGH, et per rectam
"lineam BL, et ipfam pofitione datam, femper exiftens [fc. ipfa HL] parallela
"fubjecto plano; ad lineam igitur eft punctum I, [fc. ad communem fectionem
"fuperficiei a recta HL defcriptæ, cum plano prædicto quod per BC tranfit]
"ergo eft punctum E ad lineam [quæ fc. formatur ducendo perpendiculares
"a linea in qua verfatur punctum I, ad fubjectum planum.] Hoc quidem
"univerfe refolutum eft, [fc. quæcunque fuerit data ratio EF ad DC.] Si
"autem hæc ratio æqualis fuerit rationi quam habet BA ad circumferentiam
"quadrantis ADC, prædicta linea [in qua eft punctum E] quadratrix efficitur."

The 29th Prop. next following is, " Poteft etiam illud per lineam fpiralem
in plano defcriptam refolvi fimili ratione."—It is unneceffary in this place
to give the inveftigation, but I fhall remark fome of the moft material
corrections of it by Dr. Simson.
In line 6. fol. 59. a. Command. " in fuperficie cylindrica, in qua eft linea
"fpiralis."—Dr. Simson fuppofes from Note F on this Propofition, that in
the MS. ufed by Commandine the reading was ἐν κυλινδροιίδει ἄρα ἐπιφανεία,
which is more correct, and is alfo confirmed by the MS. Bull. It fhould
therefore have been tranflated " in fuperficie cylindroeide."—Alfo, inftead

of " in qua eft linea fpirali," it fhould be " quœ formata eft a linea fpirali:" for in MS. Bull it is " τῇ ἀπὸ τῆς ἱλικος," which is plainly the true reading.

Lin. 8. " ad lineam eft ipfum K," add, " fc. ad eam lineam quæ eft " communis fectio fuperficiei cylindroeides et fuperficiei conicæ." Lin. 11. for " ergo K et I funt in fuperficie," read, " ergo eft punctum I in fuperficie, fc. " in ea quam vocavit πληκλειδος." Lin. 13. for, " ad rectam igitur lineam eft " punctum I;" read, " ad lineam igitur eft punctum I,* " fc. eam quæ com- " munis eft fectio fuperficiei πληκλειδιος† cum plano quod eft fuper rectam BC, " et ad fubjectum planum eft inclinatum per dimidium recti."

Note H. p. 65.

For the fatisfaction of fome readers, I fhall infert Dr. Simson's Paper on Series, enclofed in his letter to Dr. Jurin, of Feb. 1, 1723, (for which fee note p. 65.) The two Preliminary propofitions are very eafy, and are now to be found in all the compleat treatifes of trigonometry; but as they are fhort, they are inferted as they ftood in Dr. Simson's original paper.

Dr. Simson about a month after received Dr. Jurin's anfwer, (of March 5, 1723,) communicating Mr. Machin's feriefes; and to his paper he has added a note, ftating the correfpondence of five of his own feriefes with thofe of Mr. Machin's. Mr. Machin communicated feven; and two of them, which had not before occurred to Dr. Simson, were readily inveftigated by him.

" Prop I. Fig. 11. Sint in circulo cujus centrum eft C duo quilibet arcus AB, " AD, qui fimul fumpti funt quadrante BL minores; erit exceffus quadrati quod " fit a femidiametro fupra rectangulum fub tangentibus arcuum AB, AD ad " quadratum a femidiametro, ut tangentes arcuum AB, AD fimul fumpti, ad " tangentem fummæ arcuum.

" Sint enim AE, AF tangentes arcuum AB, AD, et fit BG, tangens fummæ " ipfarum BD, et juncta CL, occurrat ipfi AF in H, ducatur vero FK parallela " ipfi AC, et occurrens EC in K, et jungatur KH. Igitur quoniam anguli " HCK, HFK recti funt, erunt puncta H, F, C, K, in circulo; ergo angulus " FHK equalis eft ipfi FCE, et proinde fimilia funt triangula rectangula

† For fome obfervations on πληκλειδος, fee Appen. II.

* This is confirmed by MS. Bull, which has πρὸς γραμμῆς ἄρα τὸ ι,.

P

" HFK, CBG. Eſt autem ratio EF ad BG compoſita ex rationibus EF ad FK
" et FK ad BG ; i. e. ex rationibus EA ad AC, et FH ad CB ſeu AC ; ergo EF
" eſt ad BG ut rectangulum EA in FH, ad quadratum ex AC; ſed eſt EA in FH
" exceſſus rectanguli EAH, i. e. quadrati ex AC, ſupra rectangulum contentum-
" tangentibus EA, AF, et eſt hic exceſſus ad quadratum ex AC ut EF ad BG.
" 2. E. D.

" Cor. 1. Hinc ſi arcus AB, AD, æquales fuerint, erit exceſſus quadrati a·
" ſemidiametro, ſupra quadratum tangentis arcus cujusvis oſtante circumferentiæ
" minoris ad quadratum a ſemidiametro, ut tangens duplicatus ad tangentem·
" arcus duplicati.

" Cor. 2. Hinc etiam inveniri poteſt tangens arcus qui multiplex eſt
" arcus alterius, cujus tangens ḋatur. Si: a tangens arcus A, et erit tangens·
" arcus n × A.

$$na - \frac{\dot{n} . n - 1 . n - 2}{1. \quad 2. \quad 3.} a^3 + \frac{\dot{n} . \dot{n} - 1 . n - 2 . n - 3 . n - 4}{1. \quad 2. \quad 3. \quad 4. \quad 5.} a^5 \ \&c.$$

$$1 - \frac{n . n - 1}{1. \quad 2.} a^2 + \frac{n . n - 1 . n - 2 . n - 3}{1. \quad 2. \quad 3. \quad 4.} a^4 \ \&c.$$

" Numeratoris hujus ſeriei ſi n ſit numerus par, ſumendi ſunt tot termini
" quot ſunt unitates in ½ n; denominatoris vero tot quot ſunt unitates in ½ n + 1.
" Si vero fuerit n numerus impar, tam numeratoris quam denominatoris,
" ſumendi ſunt termini $\frac{n+1}{2}$. Hanc ſeriem video apud Christ. Wolfium,
" ex ſinubus derivatum.

" Prop. 2. fig. 12. Sint duo quilibet arcus AB, AD, quorum major non
" excedit quadrantem; erit quadratum a ſemidiametro una cum rectangulo
" ſub tangentibus arcuum ad quadratum ex ſemidiametro, ut diſſerentia tangen-
" tium ad tangentem differentiæ arcuum.

" Simili prorſus modo quo precedens demonſtratur.

" Cor. 1. Hinc ſi arcus minor AB, æqualis fuerit dimidio quadrantis, erit
" radius una cum tangente majoris arcus ad ipſorum differentiam, ut radius ad
" tangentem arcus quo AD ſuperat dimidium quadrantis. Si vero arcus major
" AD æqualis fuerit dimidio quadrantis; erit radius una cum tangente minoris·
" arcus, ad ipſorum differentiam, ut radius ad tangentem arcus quo dimidium
" quadrantis ſuperat arcum AB, i. e. in utroque caſu erit ſumma tangentium ad
" ipſorum differentiam ut radius ad tangentem differentiæ arcuum. Corolla-
" rium vero hoc, ni fallor, jamdudum innotuit.

" Præcedentium ope innumeræ feries exhiberi poffunt ad longitudinem cir-
" cumferentiæ inveniendam. Plurimæ vero, ad ufum paratiores, in fequente
" ferie continentur. Sc. fint tangentes $\frac{1}{a}$, $\frac{1}{2^n a}$ et fit A arcus cujus tangens eft $\frac{1}{a}$
" et erit, exiftente radio $= 1$,

$$\left.\begin{array}{l} \frac{1}{a} - 1 \times \overline{\beta + 2\gamma + 4\delta + 8\epsilon} \ \&c. \\[4pt] -\frac{1}{3} \times \frac{2^n}{2^n a^3} + \frac{1}{3} \times \overline{\beta^3 + 2\gamma^3 + 4\delta^3 + 8\epsilon^3} \ \&c. \\[4pt] +\frac{1}{5} \times \frac{2^n}{2^n a^5} - \frac{1}{5} \times \overline{\beta^5 \times 2\gamma^5 \times 4\delta^5 \times 8\epsilon^5} \ \&c. \\[4pt] -\frac{1}{7} \ \&c. \end{array}\right\} = \text{Arcui A.}$$

$$\left.\begin{array}{l} \text{Eft vero } \beta = \dfrac{1}{\overline{4aa \times 3} \times a} \\[8pt] \gamma = \dfrac{1}{\overline{16aa + 3} \times 2a} \\[8pt] \delta = \dfrac{1}{\overline{64aa + 3} \times 4\beta} \\[8pt] \epsilon = \dfrac{1}{\overline{256aa + 3} \times 8a} \\[4pt] \qquad \&c. \end{array}\right\} \begin{array}{l} \text{Terminorum autem } \beta, \\ \gamma, \delta, \&c. \text{ tot fumendi} \\ \text{funt, quot funt uni-} \\ \text{tates in numero in-} \\ \text{tegro } n. \end{array}$$

" Facile deducitur hæc feries ex Cor. 1 et 2 Prop. 1. et ex Prop. 2. Sit jam
" arcus D datum habens tangentem, datamque rationem ad circumferentiam ;
" Sc. fit 2 r D=circumf. =c; fitque mA is arcus qui vel proxime excedit arcum
" D, vel ab ipfo proxime deficit exiftente m numero integro; et inveniatur
" tangens arcus mA ope Cor. 2. Prop. 1. deinde inveniatur tangens differentiæ
" arcuum mA & D ope Cor. 1. Prop. 2. Sit tangens hic t, et erit arcus cujus
" eft tangens, $t - \frac{1}{3} t^3 + \frac{1}{5} t^5$ &c. ut notum eft ; vocetur hic arcus B, et erit
" $mA \overline{+} B = D$, unde $rmA \overline{+} rB = rD =$ circumferentiæ circuli cujus diameter
" =1.

Ex. 1. Sit $a=4$, $n=o$, $r=4$, $m=3$, eritque $t=\dfrac{5}{99}$.

Et $\quad rmA + rB = \dfrac{12}{4} + \dfrac{20}{99}$

$$- \frac{1}{3} \times \frac{\dfrac{12}{4^3} + \dfrac{20^3}{99^3}}{}$$

$$+ \frac{1}{5} \times \frac{\dfrac{12}{4^5} + \dfrac{20^5}{99^5}}{}$$

&c.*

N. B. Quoniam $\dfrac{1}{4}$ eft tangens arcus A erit tangens arcus m A. *i. e.* 3 A $=$ $\dfrac{47}{5^2}$ unde tangens arcus $45° - 3$ A erit $\dfrac{5}{99}$.

Ex. 2. Sit $a=5$, $n=1$, $r=4$, $m=4$; eritque $\beta = \dfrac{1}{515}$; $t = 239$.

Et $\quad rmA - rB = \dfrac{16}{5} - \dfrac{16}{515} - \dfrac{4}{239}$

$$- \frac{1}{3} \times \frac{\dfrac{32}{10^3} - \dfrac{16}{515^3} - \dfrac{4}{239^3}}{}$$

$$+ \frac{1}{5} \times \frac{\dfrac{32}{10^5} - \dfrac{16}{515^5} - \dfrac{4}{239^5}}{}$$

&c. &c. &c.

Ex. 3. Sit $n=o$ manentibus cæteris.

Tunc $\quad rmA - r B = \dfrac{16}{5} - \dfrac{4}{239}$

$$- \frac{1}{3} \times \frac{\dfrac{16}{5^3} - \dfrac{4}{239^3}}{}$$

$$+ \frac{1}{5} \times \frac{\dfrac{16}{5^5} - \dfrac{4}{239^5}}{} \dagger$$

" Et manifeftum eft feries has celerius convergere quo major eft n, terminos " vero ipfarum, propter eandem caufam, complexiores evadere, et proinde plus " laboris ad ipfos computandos requiri, ut ex duabus ultimo præcedentibus " feriebus liquet.

* " This is Mr. MACHIN's 4th, in the paper he fent me down in Dr. JURIN's Letter of " the 5th of March 1723."

† " Quæ eft feries optima a doctiffimo Dom. MACHIN inventa." Et eft ejus 6ta.

Ex. 4.　Sit $a=2$, $n=o$, $r=4$, $m=2$. Eritque $t=\beta=\dfrac{1}{7}$

$$\text{Et } rmA - rB = \frac{8}{2} - \frac{4}{7}$$

$$-\frac{1}{3} \times \overline{\frac{8}{2^3} - \frac{4}{7^3}}$$

$$\times \frac{1}{5} \times \&\text{c.}$$

This is Mr. MACHIN's 2d.

Ex 5.　　Sit $a=3$. Cæteris manentibus.

$$\text{Eritque } rmA + rB = \frac{8}{3} + \frac{4}{7}$$

$$-\frac{1}{3} \times \overline{\frac{8}{2^3} + \frac{4}{7^3}}$$

$$+\frac{1}{5} \times \&\text{c.}$$

This is Mr. MACHIN's 3d.

Ex. 6.　Denique, fit $a = 2$, $n = o$, $r = 4$, $m = 1$, eritque $t = \dfrac{1}{3}$.

$$\text{Et } rA+rB = \frac{4}{2} + \frac{4}{3} - \frac{1}{3} \times \overline{\frac{4}{2^3} + \frac{4}{3^3}} + \frac{1}{5} \times \overline{\frac{4}{2^5} + \frac{4}{3^5}} \&\text{c.}$$

Quæ omnium eft fimpliciffima. Eft 1ᵐᵃ Dom. MACHIN.

Similiter pofito $r=6$, derivantur aliæ feries, ut fi fit

$a=\sqrt{3}$, $n=1$, $m=2$. erit $t = \dfrac{1}{15\sqrt{3}}$.

$$\text{Et } rmA - rB = 2\sqrt{3} - \frac{2\sqrt{3}}{5 \times 3}$$

$$-\frac{1}{3} \times \overline{\frac{\sqrt{3}}{2 \times 3} - \frac{2\sqrt{3}}{5^3 \times 3^4}}$$

$$+\frac{1}{5} \times \overline{\frac{\sqrt{3}}{2^3 \times 3^2} - \frac{2\sqrt{3}}{5^5 \times 3^7}}$$

$$\&\text{c.}$$

" Convergit hoc duplo celerius ferie fimplici ex tangente: $\frac{1}{\sqrt{3}}$. derivata, fed plus
" quam duplo labore."*

Thefe few notes were added by Dr. Simson to his original paper, after he
received Mr. Machin's feriefes. ....

Above thirty years after this correfpondence, Dr. Simson gives fome account
of it to the late Earl Stanhope, in a letter dated Jan. 9, 1758, of which the
following extract will not be unacceptable.

After acknowledging fome ingenious communications of feries from Lord
Stanhope, he adds: " In the beginning of the year 1723 I fent up to Dr.
" Jurin, Secretary to the Royal Society, a method of deriving feries for finding
" the circumference of the circle from the feries. $1 - \frac{1}{3} + \frac{1}{5}$ &c. which by itfelf
" is of no ufe. I had before this defired him to inform me if he knew if any
" application had been of this feries for finding of others that converged quickly;
" he wrote to me that he knew of none, and defired me to fend up the paper,
" which I did, and gave feven feries as examples of the method. He wrote me
" in anfwer, that on communicating it, he found that Mr. Machin had in the
" year 1705 or 1706, given in a paper to the Royal Society upon this affair,
" which he afterwards took back again; but that he had prevailed on Mr.

* " In the paper wrote and figned by Mr. Machin, which Dr. Jurin fent me inclofed
" in his letter of the 5th of March 1723, there are feven feriefes fet down for finding the
" arch of 45°, the
    " 1ft of which is the fame with the 6th in this letter; and the
    " 2d is the fame with the 4th of this; and the
    " 3d is the fame with the 5th of this; and the
    " 4th is the fame with the 1ft of this; and the
    " 6th the fame that is here the 3d.
" Mr. Machin's 5th is for the arch of 45°. as all his others are, when radius is 1.

$$\frac{3}{4} + \frac{1}{20} + \frac{1}{1985} - \frac{1}{3} \times \overline{\frac{3}{4^3} + \frac{1}{20} + \frac{1}{1985}} + \frac{1}{5} \times \text{&c.}$$

" His 7th is $\dfrac{8}{10} - \dfrac{1}{100} - \dfrac{11}{5637} + \dfrac{\overline{10893}}{4148022056636}$

$$- \frac{1}{3} \times \frac{8}{10^3} - \frac{1}{100^3} - \frac{11^3}{5637^3} + \frac{\overline{10893}^3}{4148022056636^3} + \text{&c.}$$

" Vide the inveftigation of the 5th and 7th feries (of Mr. Machin) in a MS. book."

" MACHIN to fend me a fhort account of it, which he did in a paper figned'
" by him, which Dr. JURIN inclofed in his letter to me. It contained feven
" feries, two of which were different from thofe I had fent, the reft were the
" fame. In cafe your Lordfhip has not feen thefe feries, I fhall fet down two.
" of them."

Dr. SIMSON then ftates the feries of Mr. MACHIN, No. 5, in the preceding
note ; and alfo one which he had fent to Dr. JURIN, and was not returned by
Mr. MACHIN, viz. Ex. 2d. in the preceding paper. Of this laft he obferves,
" This feries is one of mine, and converges very quickly ; and becaufe one part
" of it confifts of powers of $\frac{1}{10}$, is raifed with little more trouble than Mr.
" MACHIN's, by the powers of $\frac{1}{5}$ and $\frac{1}{239}$, which was publifhed in Mr.
" JONES's *Synopsis*. It gives the circumference to the diameter 1." He adds,
" The feries which Mr. EULER gives in page 107, vol. i. of his *Introductio in*
" *Analysin Infinitorum*, (Laufan. 1748,) is one of thofe which I fent up, and
" which Mr. MACHIN fent down, [being Ex. 6. before mentioned] but it does
" not converge fo quickly as the foregoing two." &c.

In the remainder of this long letter are fome other obfervations on this and
fome other matters in the work of EULER referred to. But the infertion of
them is unneceffary, fince of thefe methods of fquaring the circle by Mr.
EULER, there is a full and explanatory detail, with many judicious remarks and
improvements by Mr. BARON MASERES, in his great repofitory of curious
mathematical tracts, the *Scriptores Logarithmici*, vol. iii. p. 169, &c.†

† It may be proper to refer the reader to feveral notices of thefe feriefes of Mr. MACHIN's,
in Baron MASERES's *Effay on the Negative Sign* ; alfo in his *Scriptores Logarithmici*, vol. iii.
p. 157 ; and in Dr, HUTTON's *Menfuration*. Dr. HUTTON's explanation of Mr. MACHIN's
feries, publifhed by Mr. JONES, is reprinted in the fame volume of *Script. Log*. with another
tract alfo of the Doctor's on the fame fubject, p. 207, which had been publifhed in the
Philofophical Transactions, and contains fome curious difcuffions refpecting fuch feriefes; and
fome important improvements.

Note I. p. 69.

In a letter to Earl STANHOPE, of 22d of March 1751, he mentions very particularly the decline of his memory, and his inability to attempt some things recommended to him by his Lordship. " Perfons of my age (now paft fixty-" three) generally lofe the ability they had when younger, of a quick and ready " imagination; and their memory, (which, in my opinion, is either the imagi-" nation of fenfations paft, or the recalling imaginations we had formerly) " manifeftly decays; and fo far with me, that I have oftentimes difficulty to " recall thofe I had the laft hour, or even a few minutes before. And in long " inveftigations, where it is neceffary to look back a good way, this inability is " moft eafily obferved, efpecially when moft of the fteps are not wrote down; for I " remember fince I could go through a longer feries of fteps without writing, " than I can now well do with the help of it.* This, my Lord, makes me " afraid that I fhall not be able to engage in the undertaking you are pleafed to " recommend to me, and which, indeed, would be very agreeable to me; the " applying the method of the ancients to the modern inventions, fo as they " might be demonftrated in fuch a way as would (to ufe your Lordfhip's juft " and elegant defcription of accuracy and ftri&nefs) convince an EUCLID, an " ARCHIMEDES, or an APOLLONIUS, rifen from the grave, to examine " them. My fcholars, Mr. MOOR, Mr. WILLIAMSON, and particularly " Mr. STEWART of Edinburgh, I hope may be able to do fomething this " way; and I fhall not fail to recommend it to them, and direct them as far " as I can."————" I hope, if GOD grant me health, I may do fomething " towards reftoring fome pieces of the ancients; the firft fix, and the eleventh " and twelfth books of EUCLID's *Elements* are near ready, excepting the " figures. Next to this, I have in moft forwardnefs APOLLONIUS *de Sectione* " *Determinata*; but PAPPUS, though I have done a good deal towards reftoring " him, is fo large, that it frightens me to meddle with him."————Towards the end of this letter he adds, " And I hope to do EUCLID juftice, if it pleafe " GOD to give me opportunity to publifh the eight books of his *Elements*, " which want of fecurity from being pirated does yet delay."

---

* I fhall mention in this place an obfervation of Dr. SIMSON on mathematical enquiry, which, though not immediately connected with the fubject of this letter, may gratify the

Note K. p. 72.

Among the fmall remains of Dr. SIMSON's mathematical correfpondence which have been preferved, are two letters from the late GEORGE LEWIS SCOTT, efq; Commiffioner of Excife, with copies of two from Dr. SIMSON in reply. The general object of thefe letters was mentioned in the memoir, and for the fatisfaction of mathematical readers, the portions of them refpecting the comparative merits of the ancient and modern analyfis are fubjoined. Mr. SCOTT's firft letter is dated London, 12th April 1764, and after fome obfervations on Dr. SIMSON's edition of the *Data*, which the Doctor had folicited from him, and in which he difcovers fome prejudice againft that book, thinking its ufe to be rather logical than mathematical, he adds:

" To be more particular, I obferved that the 27 firft Propofitions relating
" to magnitudes and their ratios merely, might be much more concifely treated,
" and receive an additional evidence from the algebraic method. I fhall give
" but one inftance from Prop. xvii. which is the xith in BARROW, and is
" by him left in darknefs; you have given it light; but would it not be fhorter
" and clearer to ftate it and its demonftration in fymbols?

" The firft hypothefis, that $x - a : y :: m : n$ where a, m, and n, are data.
" The thefis that $x - $ dat$: x + y$, in ratione data. The hypothefis may be
" changed into

" $x - a : y :: a : b$ therefore $x - a + y : x - a :: a + b : a :: a : \dfrac{a^2}{a+b}$.

" Therefore by 12 . 5th Elem. $x - a + y : x - a :: x + y : x - a + \dfrac{aa}{a+b}$

" and $x - a + y : x - a :: a + b : a$

" Therefore $x - a + \dfrac{a^2}{a+b} : x + y :: a : a + b$. *q. d. e.*

" The fecond hypothefis is that $x - a : x + y :: a : b$. which may be fhewn in
. " like manner.

reader. It is alfo in a letter to Earl STANHOPE, April 9, 1758. " What your Lordfhip
" fays of the ufefulnefs, and often the neceffity, of ufing the method of induction, I have
" frequently had the experience of; and though it be a proof of the weaknefs of the human
" mind, it is at the fame time a good help to the finding out whether a Propofition be true or
" not, and moft powerfully excites us to fearch after a ftrict demonftration."

" I have followed your reafoning ; mine is effentially the fame; but is not
" the diftinction by fymbols of the indeterminate from the determinate or
" given quantities or magnitudes concerned in the queftion of ufe? and is not
" the concifenefs of the expreffion of great ufe alfo? Many of the Propofitions
" of the data might be thereby reduced to intuitive evidence. In this particular
" demonftration, it is fhewn at once, what magnitude is to be taken from x
" (viz. $a - \frac{a^2}{a+b}$), that the remainder may be in a given ratio to $x+y$. I men-
" tion thefe particulars only to introduce a more general obfervation, which is,
" that wherever magnitudes and their ratios are confidered, the method of the
" ancients feems inferior to that of the moderns, with refpect to facility, though
" not at all fo in point of accuracy.   When I look into APOLLONIUS, I
" admire the great man; and I may fay the fame of GREGORY of St. Vincent
" and of HUYGENS; but who can help complaining of the tedioufnefs of their
" demonftrations? They feem to me in many cafes like a man who would
" reject the Indian fymbols of number, and perform all the arithmetical
" operations in words at length. This might be done; but to what good
" purpofe? When I fay to what purpofe, I mean geometrical purpofe. Other-
" wife I much approve of the elements being demonftrated in words at
" length, becaufe of their logical ufe. For in the affairs of life we have no
" fymbols, but common words; and it is of fervice to habituate the mind, to
" accurate and diftinct expreffions, and even to formal fyllogifms."

     In the fame letter after, mentioning the Propofitions of the *Data* from
the 27th to the 62d, as very eafy, he adds: " In thefe Propofitions I fee
" little or no room for algebra; and indeed, in queftions of pofition, it often
" comes in aukwardly enough. Dr. WALLIS, one of the great patrons of
" this method, owns it. It is therefore a proper enquiry to determine, if
" poffible, when one method is to be ufed, and when the other, You feem to
" promife fomething of this kind in the preface to your *Conics*. I wifh you
" could find refolution to execute what I am perfuaded is almoft ready in your
" manufcripts. THOMAS SIMPSON, in his *Exercifes*, fays, that fometimes
" the algebraical, fometimes the geometric method is preferable. I believe it.
" But it feems to me that notwithftanding the praifes beftowed by Sir ISAAC
" NEWTON and his followers on the method of the ancients, his own *Principia*
" are but algebra difguifed, as MACHIN ufed to fay. Though at the fame

" time he owned that he thought Sir Isaac's tafte of demonftration preferable
" to Huygens's. I own myfelf of his opinion, as the only difficulty attending
" Sir Isaac's arguments, are the fteps fuppreffed for the fake of brevity."

I fhall add an extract from another letter of Mr. Scott's, dated 1ft May
1764, as it is connected with the fubject of the preceding.

. . When mentioning Dr. Simson's edition of the *Data*, he adds, " I think
" that fuch as would form themfelves upon the model of the ancients, will
" do well to ftudy the *Data*. I even wifh that I had been initiated in that way.
" But having once begun in the modern analyfis, I have not had leifure to
" ftrike into another road, efpecially, as, to confefs the truth, it feems to me
" much more tedious and more doubtful with refpect to fuccefs. I mean
" not that the method of the ancients leaves any doubt in the mind; but
" that the fuccefs of attempting the folution of a difficult problem is more
" doubtful in the way of the ancients, than in that of the moderns; although,
" when the folution is once found, it be often true that demonftrations formed
" according to the ancients are more agreeable, and more fatisfactory to
" the mind, than thofe derived from algebra. I fay often, for it is far from
" being always fo. To me a demonftration derived from the confideration
" of a multitude of ratios, and their compofitions, and expreffed in the
" manner of the ancients, is fo far from being clearer than the algebraical
" method, that it feems vaftly more difficult and obfcure. And I would
" appeal to any beginner, and afk him, whether the general algebraic folution
" of the problem for determining the foci of Lenfes, or Huygens's method,
" be moft clear. I might bring many other inftances, where Propofitions
" can be demonftrated in the fame manner as Euclid has done in his
" firft four books, which often reduce the truth of a Propofition to almoft
" intuitive evidence. I admit that it would be in vain to look out for any
" affiftance from algebra; and yet I think with Dr. Barrow, that moft
" of the Propofitions of the fecond book are rather clearer when treated
" algebraically, than in Euclid's method. The perfpicuity of the fourth
" of that book has often been infifted on. It is very clear, undoubtedly;
" but ftill our friend Dr. Matthew Stewart thinks that the original
" demonftration was from the preceding Propofitions, without a figure,
" which is entirely analogous to the algebraic method. The fifth book is
" not peculiar to geometry, but is equally applicable to quantity as it is to

" magnitude; nor does the doctrine of proportion receive any evidence from
" its being expounded by lines. The symbols of algebra would, I think, be
" of ufe here. In the fixth book there are many Propofitions which may
" be treated either geometrically or algebraically, without any preference of
" advantage on either fide: for wherever an analogy is ufed, it may as juftly
" be faid to be algebraical, as it can be called geometrical. I may demonftrate
" geometrically the 47. El. 1. from the doctrine of fimilar triangles, and I
" may do the fame algebraically. The two demonftrations differ only in ftile;
" and GUISNEE is miftaken, when he fays this theorem cannot be demonftrated
" algebraically. I readily admit that the theorems about the congruity and
" fimilitude of triangles cannot be demonftrated by algebra, and that thefe
" are the true geometrical elements ; which being eftablifhed, the other Pro-
" pofitions in the *Elements* may be derived, either by the method of the
" ancients, or by that of the moderns, equally well ; at leaft, if the ancient
" method has the advantage in eafy queftions, that of the moderns feems
" of almoft indifpenfible neceffity in more difficult enquiries. I admit, however,
" that many parts of your writings, and of Dr. MATTHEW STEWART's, might
" furnifh ftrong objections to what I advance, nor do I pretend to anfwer
" fuch objections: all I fay at prefent is, that I know a perfon very well verfed
" in both ancient and modern geometry, who feems diffident of his own
" abilities to comprehend Dr. STEWART's determination of the Sun's diftance.
" He and I would, I believe, find the fame difficulties in Dr. STEWART,
" that BULLIALDUS did in ARCHIMEDES. And to conclude the fubject,
" I muft confefs that I cannot find that fuperiority of evidence in the method
" of the ancients, that you and many other able geometers, in this ifland,
" contend for. Their great genius and accuracy I do fee, and I fee fometimes
" the precipitation and innacuracy of modern geometers. But thefe are the
" faults of individuals, and not of the art."

In the fame letter, after fome remarks on the data not connected with the
particular fubject of the preceding extract, he adds, " In your notes you
" bring two inftances of the ufe of the data, and think that the conftruction
" of the laft problem could not be derived from an algebraical folution.
" I find, indeed, that the algebraic folution is more complicated than one
" would at firft fufpect, on reading the problem. Even in very fmall numbers,
" the folution brings you to large ones. Thus, if $x : y : : 6 : 7$. alfo $y - 1 : z : : 3 : 4$

" and $xx + zz = 100$. I find $x = \dfrac{168}{277} + \dfrac{\sqrt{28224 + 2203812}}{277}$ and

" $\dfrac{\sqrt{2232036}}{277} = \dfrac{1494}{277}$, therefore $x = \dfrac{168}{277} + \dfrac{1494}{277} = \dfrac{1662}{277} = 6.$

" and $x = -\dfrac{1326}{277}$, therefore $y = 7$, and $-\dfrac{1547}{277}$

Alfo $z = 8$, and $-\dfrac{2432}{277}$, and both roots anfwer the

" queftion.

" In the general folution, fuppofing $\quad x : y :: a : b$
$$y - a : z :: a : c$$
$$xx + z^2 = d^2$$

" I find $\dfrac{bc}{a^2} x - c = z$, then taking $a : b :: c : f = \dfrac{bc}{a}$ and

" $a : \sqrt{aa + ff} :: \sqrt{aa + ff} : b = \dfrac{aa + ff}{a}$. Alfo $b : c :: f : \dfrac{cf}{b} = p$. I have

" $xx - 2px = \dfrac{a}{h}(dd - cc)$ and the conftruction derived by this method is not

" fo fimple as yours. But the equation $xx + zz = dd$ being obvioufly a locus

" ad circulum, and the equation $\dfrac{bc}{aa} x - c = z$, a locus ad rectam, I find an

" eafy conftruction by combining them. For taking $a : b :: c : f = \dfrac{bc}{a}$ I have

" $\dfrac{f}{a} x - c = z$, which is very eafily conftructed; and then defcribing a circle

" from the origin of the abfciffa x with the Radius d, I have the other Locus,

" the interfection of which with the former determines the value of x. This

" is not lefs fimple than yours; and you have alfo taken the advantage of the

" Locus ad circulum.——Hitherto there appears not to me any advantage in

" the ancient method. Nor do I fee any from the firft of the inftances given

" at the end of your *Conics*. It is true your folution is more fimple than

" NEWTON's, but your inveftigation is not peculiar to ancient geometry. For

" what fhould hinder an algebraift from fuppofing the angle BCD (in your figure)

" equal to BAC, and then calling BD = x, DC = y, and AB = a, we have by

" the fimilar triangles, and by multiplying the extremes and means of the

" analogy, $ax + x^2 = y^2$, which is a Locus ad hyperbolam æquilateram, very

" easily conftructed. If Sir Isaac's inveftigation be more complex, it is not
" becaufe he made ufe of algebra, but becaufe he did not make a convenient
" election of terms. Now this election cannot be taught precifely, though
" Sir Isaac himfelf has given many excellent obfervations relating thereto.
" It is a fagacity and habit to be acquired by the ftudy of good models; and in
" this refpect there can be no doubt but the ftudy of the ancients may be of
" great ufe, even to an algebraift. I believe Mr. D'Alambert fpeaks truly,
" when he fays, that the ftudy of geometry is perhaps too much neglected."
    " After fome other obfervations, he adds. " To conclude, I long much to
" fee a treatife of which you gave us hopes in the preface to your Conics, when
" you fay in quibus autem differat analysis geometrica ab ea quæ calculo
" instituitur algebraico, atque ubi hæc aut illa sit usurpanda alias disserendum.
" Such a treatife is much wanted; and I am perfuaded that you muft have it
" almoft ready for the prefs among your papers, fo that the labour of publifhing
" would not be great."

    Thefe letters, it is obvious, were written by Mr. Scott without any inter-
mediate letter from Dr. Simson. I fhall fubjoin the material parts of the
Doctor's two anfwers, which he difpatched feparately, though he mentions
that he had received the fecond before he had finifhed his anfwer to the firft.

Dr. Simson's firft letter, difpatched July 27, 1764.

    " I am now, by the weaknefs of my memory and of my eyes, in no good
" condition to write letters of any length; but yet I fhall try to anfwer the
" feveral particulars you mention, to the beft of my ability. I am glad the
" Problems have come to your hands. You perhaps have not yet leifure to
" read them, elfe I would have expected your opinion about them. Your
" cenfures and corrections I value more than commendations. In the preface
" to the Data, it is faid, that they are of the moft general and neceffary ufe
" in the folution of problems of every kind; and whoever tries to inveftigate
" the folutions of problems geometrically, will foon find this to be true; for
" the analyfis of a problem requires that confequences be drawn from the
" things that are given, until the thing that is fought be fhewn to be given.
" Now fuppofing that the book of the Data were not extant, thefe confequences
" muft be found out and demonftrated from the things given; which, in moft

" cafes, would take no little time and pains. But now having that book, the
" Propofitions of it need only be cited, or fuppofed to be known; for though
" the ancients did not cite the particular Propofitions either of EUCLID's
" *Elements,* or of his *Data,* yet they fuppofed them to be known, as is evident
" from the folutions of problems given by PAPPUS, in which he always makes
" ufe of the Data as being known to every geometer. The Elements are
" neceffary in every geometric reafoning, whether in demonftrating theorems
" or in refolving problems; and in the analyfis of problems the Data are of
" as general and neceffary ufe."

" The demonftration of the 17th Dat. which you gave in fymbols, is very
" right; the fhortnefs of the expreffion by fymbols is no doubt of ufe; but
" then there is greater hazard of committing miftakes by ufing them, than
" when the words at length are made ufe of; and I believe that Dr. BARROW's
" miftake in this Propofition has been owing to this fhort-hand writing, which
" coft more time to underftand, than if he had ufed the ordinary way. As A
" is with him the given magnitude, and $\frac{B}{C}$ the given ratio, it is plain that
" A + B is the firft of the magnitudes mentioned in the enunciation; from
" which taking A, the remainder B has to C, the fecond magnitude, a given
" ratio; and the thing he fhould have demonftrated is, that A + B is major
" A + B + C data quam in ratione, which he has not done, but only fhewed
" that A + B is major B + C *d. q.* in r. Many fuch miftakes are obfervable
" in problems done algebraically; and in the 31 Prop. lib. i. of the *Loci plani*
" p. 96, I have taken notice of a remarkable one in SCHOOTEN's *Exer-*
" *citationes,* p. 249. The given magnitude AE (in your notation $a - \frac{a^2}{a+b}$,) which
" is to be taken from AB (or *x*), is eafily feen to be the excefs of AD (or *a*)
" above ED, which is the 4th proportional to DC, DB and AD the given
" magnitude; and if the given ratio of DB to BC be *a* to *b*, as you make it,
" then the ratio of DC to DB will be that of *a + b* to *a*.

" You feem to think, that wherever magnitudes and their ratios are con-
" fidered, that the algebraical method is better than that of the ancients in
" refpect of facility, though not of accuracy. But what if the conftruction
" deduced from the firft method, be nothing fo good, either in refpect of
" fhortnefs or elegance, as that from the other? which is often the cafe.

" You may at leifure try the algebraic folution of Prob. 2. in pages 465 and
" 466 of the Englifh EUCLID, which will furnifh you with an inftance of this.

" The length of the demonftrations in the method of the ancients arifes,
" for moft part, from the ftrict regard they had to accuracy; yet I do not
" fay that none of them can be made fhorter; for fometimes, by making
" ufe of a more proper medium of the demonftration, they can be fhortened :
" an inftance of which you have in the 20th Prop. b. ii. of APOLLONIUS's
" *Conics*; which, by making ufe of the properties of the Affymptotes inftead of
" thofe of the Diameters, is made fhorter in the 40th Prop. of b. iii. of the
" book which I publifhed fometime ago. I expreffed myfelf indiftinctly
" in the preface to that book, which has made fome think I defigned to
" publifh fomething, to fhew when the ancient and when the modern method
" ought to be ufed; but I meant to fay nothing more than that it was not
" proper to be done in that preface."

" I cannot guefs on what account Mr. MACHIN, to whom you affent, thought
" Sir ISAAC's tafte of demonftration preferable to that of Mr. HUYGENS.
" Though I have the greateft veneration for Sir ISAAC's profound and penetrating
" genius, and his moft valuable and ufeful difcoveries, yet I cannot but prefer
" HUYGENS's elegance and perfpicuity, efpecially in his *Horologium oscilla-*
" *torium*, to Sir ISAAC's; nor do I think the firft has fallen into fo many
" miftakes as the other, in things purely mathematical."*

I fhall fubjoin fome extracts from Dr. SIMSON's letter in anfwer to Mr.
SCOTT's 2d letter, of May 1, 1764.

" DEAR SIR,

" Your approbation of the edition of the *Data*, with which you begin your
" fecond letter of May 1ft, gives me great fatisfaction. You obferve that the
" demonftrations made by a multitude of ratios and their compofitions, ex-
" preffed in the manner of the ancients, are more difficult and obfcure than
" the algebraic method. I confefs a multitude of them makes a demonftra-
" tion difficult in any of the methods, and I found it was not only difficult to
" my fcholars to underftand the 20th Prop. of HUYGENS's *Dioptrics*, but
" uneafy to me to prelect it to them, by reafon of the great number of ratios
" that are confidered in the demonftration, where there is only a ftraight

* Sir ISAAC NEWTON himfelf entertained a high opinion of the elegance of HUYGENS's
tafte in geometry. See Pemberton's View, Preface.

" line and its parts to aid the memory: and on this account I contrived a
" demonstration by help of a figure, by which the composition of ratios is
" avoided, and nothing but the properties of parallel lines, and similar
" triangles, enter into; it which made it much easier both to my scholars and me."

" No doubt the fourth of the second book of the *Elements* might have been
" shewn from the third and the second of it, and EUCLID, or even a far less able
" geometer, could not but see it; but he preferred the way by figures in all the
" demonstrations of the second book, as shewing the equality of the spaces
" compared to the eye, which is more clear than when it is perceived by
" the imagination. As to the fifth book, I do not see that the demonstrations.
" in it can receive any help from algebra; and the straight lines made use of
" in it make the demonstrations clearer and easier, than they would be without
" them. I do not understand your meaning, when you say any analogy
" may be called algebraical as well as geometrical. The expressing lines by
" a single letter does not make analogy or any thing else algebraical, any
" more than when they are expressed by the two letters at their ends, nor do I
" think any thing can be called algebraical, where no operation peculiar to
" algebra is made use of. I should be glad to see an algebraical demonstration
" of some proposition in the 5th book, *ex. gr.* of the 17th, by which perhaps
" I might understand your meaning. The 47th Prop. of 1st B. cannot be done
" by similar triangles in the place where it now stands; and when it is done
" by them, I do not see that the demonstration has the least dependence
" on any thing in algebra. You think algebra almost indispensibly necessary
" in difficult enquiries : I have, on the contrary, been obliged to make use of the
" ancient method in many problems, in which I could not find that algebra was
" of any use to me; such were the two problems, the solutions of which I sent
" some time ago; and so are many others, of which the following is one. *Three*
" *points and three straight lines being given in position, it is required to place*
" *a triangle so that the three angular points may respectively be on the three*
" *given straight lines, and the three sides pass through the three given points."*

" The solution which you give of the last problem at the end of the *Data,*
" by help of two *Loci,* is very good; but I am not yet persuaded, that the con-

* The enunciation in this letter has reference to a figure which is here omitted. It
is also to be remarked that in the *Opera Reliqua,* p. 388, is a solution of a case of this
problem, viz. when the three given points are in one straight line.

" ſtruction ariſing from it, or the demonſtration of it, will be quite ſo ſimple
" as thoſe given in that place : if you had leiſure to write them out, without
" making uſe of any algebraical notation or operation, they might then be
" better compared. You think I have taken advantage of the *Locus ad Cir-*
" *culum*, but I aſſure you it never came into my mind : the 34th Prop. in
" the *Data*, as you will ſee by looking to it, does indeed require the deſcriptiou'
" of a circle, by which the poſition of the ſtraight line AD (in Prob. p. 465,)
" is found.

" It would take up too much time, and writing too much for my eyes, to
" ſhew the advantage of the ancient method above the algebraic, and how the
" precepts given in this laſt method, particularly in the *Arithmetica Universalis*,
" lead thoſe who obſerve them, from the right ſolution of geometrical
" problems, into ſuch as are quite out of the natural method ; many inſtances
" of this occur in that book, among which is that problem which is the
" firſt at the end of the ſecond edition of the *Conics*. You ſay the ſolution
" there given is not peculiar to the ancient geometry, and you add, 'what ſhould
" hinder an algebraiſt to ſuppoſe the angle BCD in my figure equal to BAC
" &c.?' True, an algebraiſt may give a geometrical ſolution of a problem, but it is
" not therefore an algebraic ſolution. The ſolution you give in the algebraic
" notation depends nothing on algebra, for the *Locus* $ax + x^2 = y^2$ flows from
" one of the firſt properties of the hyperbola demonſtrated in the *Conics*,
" and is the very ſame which is made uſe of in the 1ſt Prob. of the appendix.
" It may be ſaid that algebraiſts are blameable, not ſo much in making a wrong
" choice of a medium, as for hindering themſelves from making any choice
" not pointed out by their rules ; one of which almoſt conſtantly made uſe
" of by them is, to let fall a perpendicular, that by help of the 47. 1ſt Elem.
" they may get an equation, which otherwiſe they could not ſo eaſily obtain.*

* An obſervation of DES CARTES tends to confirm ſome of Dr. SIMSON's remarks in
this letter ; for which ſee BRANKER's Algebra, p. 65, (1668.) DES CARTES, in a letter
" not yet printed, writes thus: *In ſearching the ſolution of geometrical queſtions, I always*
" *make uſe of lines parallel and perpendicular as much as is poſſible ; and I conſider no other*
" *theorems but theſe two.* (The ſides of like triangles have like proportions,) *and* (in
" rectangle triangles the ſquare of the greateſt ſide is equal to the ſquare of the other two
" ſides,) *and I am not afraid to ſuppoſe many unknown quantities, that I may reduce the*
" *proposed queſtion to ſuch terms, as that it depends on no other theorems but theſe two.*" This
letter of DES CARTES, however, was afterwards printed, Amſt. 1683, Epiſt. 72, part iii.

" I , forgot, when mentioning the laft problem, at the end of the *Data*, to
" obferve that in the equation you firft bring it to, you give the value of
" what is called the negative root, viz. $x = -\dfrac{1326}{277}$.† It is one of the abfur-
" dities introduced into algebra in the laft age, to fuppofe every equation has
" as many roots as there are units in the index of its higheft power, and
" confequently that every quadratic equation has two : but the contrary, in
" the prefent cafe, can eafily be fhewn. Your equation is of the fame form
" with this $x^2 - 4x = 12$, the root of which you know is $x = \sqrt{12+4} + 2$ or 6.
" And that this equation can have no other root, *i. e.* that there can be
" no number but 6, from the fquare of which taking the quadruple of the
" number, the remainder will be 12, appears thus. Suppofe it poffible, that
" y is alfo the root of the equation, or that $yy - 4y = 12$. Then y muft be
" either greater or lefs than 6 or x; (if any deny this axiom, I will reafon
" no further with him on this matter;) fuppofe that it is greater, $y - 4$ is greater
" than $x - 4$, and therefore the product of y by $y - 4$ is greater than the product
" of x by $x - 4$; that is, $yy - 4y$ is greater than $xx - 4x$, and confequently than
" 12. Therefore $yy - 4y$ is not equal to 12 ; that is, y is not the root of the
" equation. The fame thing follows, if y be faid to be lefs than 6.

" *The passage you have set down from the preface to the Conics, might,*
" *I confess, make any one think that I designed to have published something*
" *on the subject there mentioned; but indeed, as I wrote in my last, I blun-*
" *dered in the expression, and have no papers, and never wrote any about*
" *that matter. R. S.*"*

page 296. The fentences fucceeding the foregoing extract give additional explanations
of his opinion, to which the reader is referred. I may alfo remark, that in this letter, (Ep.
72.) and in the following, (Epift. 73.) fome difficulties are mentioned in the algebraical
folution of the cafe of the *tactions*, in which three circles are given.

† Dr. SIMSON's notions about negative quantities in algebra have been frequently mentioned. His comment on the negative root of a quadratic equation, mentioned in Mr.
SCOT's letter, though not connected with the other matters in this correfpondence, is retained
as a characteriftic fpecimen, in his own words, of the Doctor's manner of treating that fubject.
He made no fecret of his opinions refpecting it : they were well known to all his
contemporaries ; and whatever objections may be made againft them, he remained firm
in maintaining them.

* This letter was difpatched to Mr. SCOT, 13th Auguft 1764 ; and no trace of correfpondence between thefe gentlemen, of a fubfequent date, remains.

With respect to this correspondence mathematical readers will form their own opinion; but it is proper to remark, that Dr. Simson, at the time of writing his letters, was in his 77th year. I venture to add a few remarks, which, to those who have not considered the subject particularly, may be acceptable.

In all those geometrical investigations, in which the chain of argument can be rendered by the ancient analysis both intelligible to the understanding, and pleasing to the imagination, Dr. Simson's reasoning in its favour will not be objected to. It must be admitted also, as he states, that unless operations peculiar to algebra be introduced, an investigation, though clothed in the characters of algebra, may yet be properly geometrical. He gives an example of the solution of a problem by the combination of two local equations; and such may always be regarded as purely geometrical, when the *Loci* employed have been demonstrated in known geometrical works, and are easily deducible from the *Data* in the problem without any algebraical operation. But even when an algebraical process is requisite for investigating the *Local Equations*; or when, in the investigation of a geometrical problem by algebra, some inferences may be more immediately drawn by means of known geometrical propositions than by the operations of that art; the algebraical solutions may by such aids be much abridged, and often rendered much more elegant.

It is further to be remarked, that when geometry comes to be employed in practical mensuration, or in physical enquiries, the most obvious and most expeditious methods of investigation, whether geometrical or algebraical, will usually be preferred. It is well known that in such applications, in general it is useful, and often absolutely necessary, to express conclusions (in whatever way they may be obtained) arithmetically; and hence it is frequently expedient to investigate by algebra such conclusions, though they might without difficulty be obtained by geometrical reasoning, as they must ultimately assume the arithmetical form in the practical uses of them. I must also observe, that Mr. Scott's remark respecting the combination of ratios is perhaps not fully answered by Dr. Simson; and that the remark may also in some cases be extended to any long series of steps, by the combination and separation of squares and rectangles arising from the segments of a straight line, by the use of the Propositions of the iid b. of Euclid, or of others founded on that book. Though a demonstration may be regarded as strictly

geometrical, when fuch ratios, and the combinations of them, are fet down
either in words at length, or in a kind of fhort-hand, by ufing fome
algebraical characters; yet when the mind lofes the diftinct perception of the
particular geometrical magnitudes compared, the evidence is fimilar in its
impreffion to that of algebraical reafoning, in which the previous demonftration
of the rules employed is the ground of our affent to the truth of the conclu-.
fions, and not the immediate perception of the geometrical magnitudes and
their relations.—An analyfis or demonftration in which many combinations
of ratios are employed, may generally indeed be much fhortened by admitting
multiplication, divifion, or other operations peculiar to arithmetic or algebra;
but the inveftigation or demonftration then becomes truly algebraical. It
muft however be admitted, in the particular cafe here fuppofed, that the
evidence of the algebraical procefs is not materially different, or rather that
the impreffion of that evidence on the mind, is not very different, from that
of the geometrical method. The fame obfervation may be extended alfo
to the cafe, when many combinations of fquares and rectangles from the
fegments of a ftraight line neceffarily make a principal part of a geometrical
analyfis or compofition.

Dr. SIMSON acknowledges the difficulty of purfuing fuch long trains of
reafoning, by combining the ratios in the Propofition of HUYGHENS particularly alluded to. He found it even expedient to contrive a new demonftration by figures, which might facilitate his own labour in lecturing, and
more effectually keep alive the attention of his fcholars, by prefenting objects
better calculated for that purpofe, and alfo for aiding the memory in retaining
the long feries of fteps in that Propofition. But in fuch Propofitions an
algebraical deduction would in many cafes be not lefs fatisfactory, and far
more expeditious, than any other method.

It muft be conceded, however, by the patrons of the algebraical analyfis, that
the ancient method, where it can be applied, with very few exceptions, is
more elegant, and gives a more immediate and more pleafing conviction
of the truth of a geometrical propofition, than the other. It muft be admitted,
alfo, that there are many geometrical problems, even in the clafs denominated
plane, of which the algebraical folution is far from being obvious, and is
often more difficult than the geometrical; even from the complexity of the
calculation, without regarding the inferiority of the conftructions which

generally refult from the folution of equations. Dr. SIMSON, in one of his letters to Mr. SCOTT, mentions an example; and others equally illuftrative of the point might eafily be pointed out.*

That many geometrical problems are more eafily refolved by algebra than by the ancient analyfis, there is no doubt ; but when the object is to obtain a folution, in which rigid accuracy is combined with perfpicuity in every ftep of both the analyfis and demonftration, the ancient method is indifputably to be preferred. But the diftinguifhing value of the modern analyfis, though not particularly confidered in this correfpondence, is the means which it affords of refolving geometrical problems, inacceffible to the ancient analyfis in its prefent ftate; and of applying geometry to phyfical inveftigations, equally beyond the reach of that more elegant, though lefs powerful inftrument. Such, indeed, has been the fuccefs of the modern analyfis, in the vaft enlargement of fome important branches of fcience, and of the arts depending on them, that all reafonings on any theoretical imperfection in its principles, which does not affect the truth of the conclufions, is fuperfeded by the wonderful extent and utility of its power and application.†

In a mifcellaneous MS. volume of the year 1722 were found, annexed to the folution of a geometrical problem, fome fhort obfervations by Dr. SIMSON on the ancient analyfis; which, though apparently fuggefted merely by that folution, and written without any particular view, yet mark the accuracy of his notions refpecting that analyfis at this early period, before he had made much progrefs in the reftoration of the loft analytical works of the ancient geometers. They may however, notwithftanding thefe circumftances, be confidered as a fuitable addition to the Doctor's letters on the fubject in his old age, which have been

---

* The problem, (117 lib. vii. PAPPI) and the extenfions of it, of which a fhort hiftory is given in note E, may alfo be mentioned as examples.

† Such is the power of the modern analyfis in many complex inveftigations, that an obfervation of WOLFIUS (tom. v. Math. Univ. p. 210) will not be confidered as altogether extravagant, even by the rational admirers of ancient geometry. " Sane fi " ARCHIMEDES et APOLLONIUS noftro ævo revivifcerent, in ftuporem raperentur vifis " inventis recentiorum, quæ per algebram fuerunt in apricum producta; neque unquam " fibi perfuaffiffent patere ad talia mortalibus aditum." The original and powerful genius of ARCHIMEDES however, had his life been prolonged, might have anticipated many of the great difcoveries of modern times; and have adorned them alfo with his elegant and accurate manner of demonftration.

given in this note. To prevent unneceffary trouble, the particular references to the problem are omitted.

" Elegans eft hoc exemplum veræ et genuinæ tam analyfios quam con-
" ftruêtionis et demonftrationis, methodo veterum inftitutarum, quafque algebra
" recentiorum nunquam præbuiffet."

" In genere notandum eft, analyfim eo puriorem et elegantiorem habendum
" effe, quo minus artificii in conftruêtione analyfi præmiffa invenitur; *i. e.* quo
" pauciores, eæque non nifi ob rationes obvias ducendæ lineæ, conftruêtionem
" iftam ingrediuntur.

" Analyfes enim hac methodo inftitutæ præferendæ funt iis quæ, quamvis bre-
" viores effe poterint, tamen ope conftruêtionis minus obviæ perficiuntur, *i. e.* hæ
" ipfæ analyfes alterâ analyfi indigent, ut ad iftas conftruêtiones perveniatur, quæ
" propterea minus perfeêtæ funt cenfendæ. Perfeêtiffimæ enim funt analyfes quæ
" a datis procedunt ad ifta quæ immediate cum iis conneêtuntur, atque ita donec
" ad quæfitum perveniatur. Verum quidem eft, eos qui fagaci et acri pollent
" ingenio, poffe per pauciores gradus ad quæfitum, quafi faltando, pervenire, fed
" hoc nullo modo poteft fieri, nifi vel animo perceperint grados omiffos, quod
" redit ad methodum defcriptam, vel tentando inciderint in conftruêtionem
" quandam, a datis fatis remotam, id quod omni cura ab analyfta eft vitandum.
" Cum nihil verâ et genuinâ analyfi magis diftat, nihil magis abhorreat, quam
" tentandi methodus; hanc enim amovere et certiffima via ad quæfitum per-
" ducere, præcipuus eft analyfeos finis; quamvis fatendum, tentando et in
" tenebris quafi palpitando, quæfitum poffe inveniri, fed hac methodo nunquam,
" nifi cogentè demum, ob deficientem analyfim, neceffitate procedendum. De
" compofitionis et demonftrationis natura, cum eæ ex analyfi femper efficiuntur,
" eodem modo judicandum.

" Porro animadvertendum eft nihil in analyfi de quæfito vel ejus quavis affec-
" tione fupponendum quod ex datis non clare fluat. *Ex gr.* In inveftigatione
" loci alicujus non fupponendum eft locum effe *rectam* vel *circulum,* quoniam
" hoc eft quod quæritur. Et hoc vitio laborat analyfis ex qua conftruêtionem
" fuam Prob. in Prop. 7, p. 23, lib. i. APOLLONII *de Locis Planis,* haufiffe
" videtur FERMATIUS, nam inveniendo duo punêta in quibus locus quæfitus
" fecat datas pofitione reêtas, et affirmando reêtam, hæc duo jungentem, effe
" locum quæfitum, videtur eum etiam in analyfi fuppofuiffe, locum quæfitum
" effe reêtam.

" Notandum denique in difficilioribus problematis tractandis multum ad
" folutionem conducere, fi a cafu aliquo fimpliciore inciperimus."

He proceeds to illuftrate this laft obfervation by a particular cafe, which is
too long for infertion.

APPENDIX.

APPENDIX I.

Account of the Mathematical Collections of PAPPUS.†

§. I. *Two first Books of* PAPPUS.

PAPPUS, an eminent Mathematician of Alexandria, flourished about the end of the fourth century.‡ In the very brief accounts which remain respecting him, he is mentioned as the author of several treatises all of which; except his *Mathematical Collections*, probably the most valuable of his writings, appear to have perished. This work, as the

† In Dr. HUTTON's learned work, the *Mathematical Dictionary*, there is an excellent abstract of the contents of the *Mathematical Collections*, to which I at first intended to have referred. But on further consideration, I found that, for the purpose of giving a satisfactory account of Dr. SIMSON's commentaries and corrections of that work, a more detailed statement was necessary, as will readily appear to the intelligent reader.

‡ See SUIDAS; (Πάππος,) also GER. VOSSIUS *de Chronologia Mathematicorum*; and MONTUCLA tom. i. In SUIDAS several works of PAPPUS are mentioned; but neither his Collections, nor a treatise alluded to by himself in b. iv. of the Collections, (COMMAND. fol. 56, 1588.)

name imports is mifcellaneous, and befides a variety of
Propofitions, both problems and theorems, contains fome
curious notices, not to be found elfewhere, of the hiftory
of mathematics and of mathematicians in his own and in
preceding times. Its chief importance however is, as has
already been mentioned, that from it alone we derive fatisfactory information refpecting the ancient geometrical
analyfis. In the preface to the feventh book, particularly,
is given a defcription of the moft valuable treatifes in the collection, which in the celebrated fchool of Alexandria* got
the title of τόπος αναλυομένος, and which was compofed by fome
of the moft eminent geometers of antiquity, for promoting the
ufe of that analyfis.

Of the eight books of the *Mathematical Collections*, the firft
and about one half of the fecond are prefumed to be loft; the
reft have reached the prefent time, though with many imperfections, and in fome paffages fo mutilated, that the meaning
cannot be certainly afcertained. The original Greek (except
fome fhort extracts) has never been printed; and the only tranflation of it, by COMMANDINE, was firft publifhed at Pifaurum
in 1588; and another edition, with little variation or improvement, was printed in 1660 at Bologna.† This tranflation is

* It is remarked by Vossius, in the before-mentioned work, " ab Euclidis tem-
" pore ufque ad tempora Saracenica, vix ullum invenire fit nobilem mathematicum,
" quin vel patria fuit Alexandrinus, vel faltem Alexandriæ dederit operam
" Mathesi."

† Though the fecond edition is faid to be corrected, yet this appears to be only in
fome prefs errors of the former edition, while new errors of the fame kind are introduced: but there is no attempt to amend the obfcure and deficient paffages in that
tranflation. From fome convenience of the bookfeller, the original edition fucces-

accompanied with a commentary, often tedious, and in fome places defective; but at the fame time extremely valuable, from the explanation which it contains of fome difficulties, and the correction of many errors in the MS. ufed by COMMANDINE, and which pervade all the MSS. of PAPPUS that have hitherto been examined. From COMMANDINE's manner of referring to the Greek, it appears that he had only one MS. for his guide. He died before the work had received his laft corrections; and no account is given of the hiftory or character of the MS. which he followed.† From a family difpute between two fons-in-law, the publication was fufpended for fome time after his death; and at length, by the munificence of his patron the DUKE of URBINO, the tranflation was printed, but confeffedly without any correction whatever of the errors or omiffions in the unfinifhed work of COMMANDINE.‡

In this ftate, however, it was a very interefting communication to the mathematicians of that age, and excited much curiofity refpecting fome branches of the ancient geometry, which, before this publication, were little confidered or known in modern Europe. Accordingly, not long after its appearance,

ffively got three title-pages and dates; the firft, Pifauri 1588; the fecond, Venetiis 1589; and the third, Pifauri 1602.

† COMMANDINE in many hundred notes quotes the CODEX GRÆCUS. In a very few places indeed he does quote the GRÆCI CODICES; but this laft expreffion feems to have arifen merely from inadvertence, as in no inftance is there any comparifon of readings of different MSS. though, in a great many cafes, from the fenfe, he corrects the blunders of the one manufcript to which he ufually refers.

‡ See the Dedication, and alfo the addrefs to the reader, in which thefe circumftances are mentioned; and it is alfo obferved that COMMANDINE had made diligent fearch for the two firft books of PAPPUS, but without fuccefs.

SNELLIUS, VIETA, MARINUS GHETALDUS, and FERMAT,
attempted, from the defcription of the τόπος ἀναλυομένος in the
feventh book, to reftore feveral of the loft books of APOLLONIUS,
which made a principal part of that collection. Afterwards
SCHOOTEN, Dr. WALLIS, and fubfequently Dr. HALLEY,
purfued the fame track; and thus a number of thefe ancient
treatifes were reftored, with various fuccefs indeed, and with
different charaЄfers of elegance and accuracy, as has already
been obferved in the foregoing Memoir.

The two firft books of PAPPUS are not in COMMANDINE's
tranflation, from their not being found in any of the MSS. to
which he had accefs; and they are wanting alfo in feveral
other MSS. remaining in various libraries of Europe.* In a
MS. of PAPPUS, in the Savilian Library at Oxford, (No. 9,)
there is preferved about one half of the fecond book, which
was publifhed by Dr. WALLIS in 1688, with a Latin tranflation,
and valuable notes, explanatory of the Greek arithmetic.

From this remaining fragment, it is reafonably conjeЄtured
by Dr. WALLIS, that thefe two books related folely to that
arithmetic; and thence he infers that the lofs of them is not
greatly to be lamented. We learn from the fragment, indeed,
that APOLLONIUS had compofed a treatife on this fubjeЄt, to
which PAPPUS frequently refers; and Dr. WALLIS derives from
it fome information refpeЄting the notation and algorithm of
the ancient arithmetic. From the defeЄtive mode of notation

* See at the end of this Appendix a fhort notice of the MSS. of PAPPUS, which
are generally known to be preferved in the libraries of Europe. It may be obferved
in this place, that fome of them, befides the Savilian MS., have the fragment of the
fecond book publifhed by Dr. WALLIS.

among the Greeks, as compared with that of modern Europe, though there be much ingenuity in fome of their methods, we need not be furprifed at the great inferiority of their fyftem. Dr. WALLIS remarks, that the bufinefs of the fecond book of PAPPUS appears to be nearly equivalent to what is now confidered as a very fimple propofition, viz. that the multiplication of any numbers, all or any of which have cyphers annexed, may be performed by multiplying thefe numbers without the cyphers, and then adding all the cyphers to the product. The firft book he with much probability conjectures to have been employed about the fimple operations of the addition and fubtraction of numbers.* And I need only further remark, that there is no appearance among Dr. SIMSON's papers of his having directed his attention to this fragment.

The fucceeding five books of PAPPUS are geometrical, and the laft is almoft entirely mechanical. They are all of a mifcellaneous character; though in fome of them there is a certain degree of method, and titles have been prefixed to them, alluding to the principal matters contained in the feveral books.† Thefe titles do not appear to have been in the MS. followed by COMMANDINE, as he does not mention them in his tranflation; and probably, indeed, they did not belong to the original work, but were added by fome commentator, perhaps of a much later age. Though I purpofe to give a particular account of the contents of the feveral books, it may be proper

* See WALSII *Fragmentum Secundi Libri* PAPPI, aud the *Epilogus*.

† The MS. BOLL. has thofe titles, from which they are copied in this Appendix. The tmo Savilian MSS. at Oxford have them, and likewife the Parifian MSS.

firſt to mention the diſtinctions of geometrical propoſitions (eſpecially of problems) affumed by the ancients, as they are ſtated by PAPPUS in two paffages of his work, expreffed nearly in the fame words.‡

COMMANDINE's tranflation of thefe paffages is given in Appendix II. and the Greek not having been printed, is added from the two Savilian MSS.§ and the MS. BULL. for the fatis-faction of thofe who may wifh to confider them accurately. From thefe extracts it appears that it was the difficulty, or rather the impoffibility, of refolving fome problems by the circle and ſtraight line, which fuggeſted the inveſtigation of other curve lines, by the defcription of which the folution of fuch problems might be accompliſhed. The doubling of the cube,* and the trifection of an arch of a circle, were two celebrated problems, which exercifed the ingenuity of the more ancient geometers, but which were found not to be refolvable by plane geometry. From the very brief accounts which remain of thefe fpecu-lations, it appears that the firſt attempt of producing new curves, which might be employed in geometrical fcience, was from the fection of a folid by a plane; and the only folids confidered in the early ſtate of the fcience, which, by fuch a

‡ The firſt paffage is in the third book, fol. 4. b. in COMMANDINE's PAPPUS, (1588,) and the other in the fourth book, fol. 60.

§ Accurate copies of thefe paffages from the two Savilian MSS. were procured for me, in the moſt obliging manner, by the Rev. Dr. ROBERTSON, Savilian Profeffor of Geometry; who alfo gave me every facility for confulting thefe MSS. and other books in the very curious collection placed under his care. I met with the fame obliging attention, when confulting fome MSS. and books in the Bodleian Library.

* Ufually called the Deliacal Problem, from the well-known tradition about doubling the cubical altar at Delos.

section could produce curves different from the circle, were the cylinder and cone. But as the sections of the latter comprehended the curves resulting from the sections of the former, the three new curves, arising from the different possible sections of the cone by a plane, obtained·the name of Conic Sections. By these curves the two before-mentioned problems were easily resolved; and from this origin, all problems requiring for their solution the description of one or more of them, were called solid, though they had no other relation to solid figures.

Some other curves were also invented by ingenious men of those times for the same purpose; but the Ancients did not pursue this branch of geometry, and confidered only a small number of such lines, without having had any notion of the unbounded number which modern speculations have brought into notice; and therefore, without proposing any principle of systematic arrangement. All curves, besides the Conic Sections, according to the account of PAPPUS, in the passages alluded to, were by the ancients denominated *Lines*, and problems resolvable only by such curves were called *Linear* Problems. All geometrical problems therefore, among them, were divided into three classes, *plane*, *solid*, and *linear*; and theorems were sometimes distinguished in the same manner, as they had reference to those three classes of lines. The superior lines treated of by PAPPUS, and other ancient writers, were the *conchoid*, the *cissoid*, the *spiral*, and the *quadratrix*; and a few others are slightly alluded to.

In the solution of the two celebrated problems, which are supposed to have given rise to the investigation of new curves, some of these last-mentioned curves have been employed, as

well as the Conic Sections. And we are informed by PAPPUS,
that the difficulty of defcribing the Conic Sections with mechanical accuracy led fome of the ancient geometers to
employ thofe higher curves, the defcription of which was found
to be more eafy. The conchoid in particular was ufed for
finding between two given ftraight lines two mean proportionals,
from which the doubling the cube was an obvious inference;
and the trifection of an arch of the circle was accomplifhed
alfo by the fame curve, and likewife by the fpiral and quadratrix.* From PAPPUS it appears, however, that the early
Mathematicians had at firft fome reluctance in admitting either
the Conic Sections or fuperior curves in the folution of problems, confidering them as not ftrictly geometrical;† but
afterwards thefe lines became objects of much curious inveftigation, even among the ancients; and in modern times ultimately were of the moft extenfive utility, both in abftract and
in phyfical fcience.

The above-mentioned paffages of PAPPUS naturally fuggeft
an obfervation which may be ftated in this place; that expreffions occur on this fubject, in the writings both of ancient and
modern geometers, in which there feems to be fome want of
that precifion and confiftency which properly belong to the
language of mathematical enquiry. The defcription of any
geometrical line from the data by which it is defined, muft
always be affumed as poffible, and is admitted as the legitimate

* Thefe problems are refolved in that manner by PAPPUS, in b. iii. and iv.

† Even fome of the more early of the Mathematicians, reckoned modern, entertained a fimilar notion; particularly VIETA.

means of a geometrical conftruction : it is therefore properly regarded as a *poftulate*. Thus the defcription of a ftraight line and of a circle are the poftulates of plane geometry, affumed by EUCLID. The defcription of the three *Conic Sections*, according to the definitions of them, muft alfo be regarded as poftulates; and though not formally ftated like thofe of EUCLID, are in truth admitted as fuch by APOLLO-NIUS,† and all other writers on this branch of geometry. The fame principle muft be extended to all fuperior lines.

It is true, however, that the properties of fuch fuperior lines may be treated of, and the defcription of them may be affumed in the folution of problems, without an actual delineation of them. A plaufible though imperfect reprefentation of geometrical lines is indeed extremely ufeful in affifting the imagination, when we are employed in the inveftigation of their affections; but the degree of accuracy with which they are exhibited is of no importance to the truth of the reafoning, or to the fatisfaction with which it is perceived. For it muft be obferved, that no lines whatever, not even the ftraight line or circle, can be truly reprefented to the fenfes according to the ftrict mathematical definitions;‡ but this by no means affects the theoretical conclufions

† See Prop. 52, 53, 54, APOLLON. b. i. in which the defcription of the three fections is affumed precifely according to the definitions given of them by APOLLONIUS.

‡ CARTESII *Geometria*, lib. ii. at the beginning. It may alfo be remarked, that even in the *Elements* of EUCLID, particularly in the xith and xiith books, certain conftructions are affumed, which, though perfectly confiftent with the rigour of geometrical demonftration, would in mechanical practice be extremely difficult. But they muft be confidered as poftulates, though not ftated to be fuch by EUCLID.

which are logically deduced from fuch definitions. It is only when geometry is applied to practice, either in menfuration, or in the arts connected with geometrical principles, that accuracy of delineation becomes important.

Among the ancients, the defcription of curve lines with a certain degree of eafe and accuracy was important; as we have reafon to believe that fuch defcriptions were often ufed by them in practical applications of geometry to menfuration and mechanics. This the limited nature of their fyftem of computation rendered particularly expedient; and influenced by this confideration, we may prefume, the Greek geometers, in the folution of problems refolvable by the *Conic Sections*, fometimes employed fuperior curves, on account of the greater facility of defcription; of which there are examples in the *Collections* of PAPPUS.* In recent times, however, the power of the modern analyfis, efpecially with the aid of trigonometry, renders the accurate delineation of curves feldom neceflary; and in the few cafes where it may be ufefully employed, thofe curves ought no doubt to be affumed, the defcription of which, in the exifting ftate of mechanical arts, moft conveniently enfures the required degree of exactnefs.

But in a fcientific confideration of the fubject, it is a principle admitted ultimately, both by the ancient and modern geometers, that the proper folutions of problems

* PAPPUS enumerates feveral folutions of the Deliacal problem, fome of them purely mechanical, and one of them by the conchoid ; a curve of a clafs fuperior to the *Conic Sections.* Sir ISAAC NEWTON even admits, that in certain cafes the conchoid may be preferred to the *Conic Sections*, on account of the fimplicity of defcription. *Arith. Univerfalis, Appendix, conftructio Æquationum linearis.*

muft be effeted by the defcription of the lowest clafs of lines by which they are practicable; and that though conftructions of problems by curves fuperior to thofe by which they are refolvable, may be demonftrably true, and in particular cafes, may be practically ufeful, yet in theory are to be rejected as irregular.

§. II. The Third Book of Pappus.

This book, as the title imports, contains geometrical problems, both plane and folid.*

The problems confidered in this book are four: firft, the duplication of the cube, or what is equivalent, the finding two mean proportionals between two given ftraight lines, which is folid : fecond, a problem refpecting the mediatates: third, a fort of paradoxical problem, to place two ftraight lines, drawn from two points in one fide of a triangle, to a point within it which may be greater than the other two fides: fourth, to defcribe the five regular folids in a fphere. Thefe three are confidered as plane.

The firft problem, which was an object of much confideration among the ancients, is refolved by Pappus in feveral of the methods then known; but he begins with a refutation of an unfcientific attempt to refolve it by plane geometry, which feems, in his time, to have excited fome attention. The

* The title in MS. Bull. is Πάππυ 'Αλιξανδρεῶς συναγωγῶτ† Γ, περιίχυ δι προβλημᾶία γεωμίίρικά ἐωίπιδα κάι ςερά.

† The tranfcript from the Parifian MS. in the Savilian library has συνχγωγῶν μιθηματικῶν.

examination of it, which had been folicited from PAPPUS
by the author and his friends, requires a long train of
reafoning, from the complex conftruction which had been
propofed by this inaccurate geometer; and fome things,
affumed by PAPPUS in this argument, are demonftrated in
four propofitions annexed to it. Both the text and the diagram have been injured by tranfcribers; but COMMANDINE's
tranflation is much amended by Dr. SIMSON, and a few of
his corrections, indeed, are confirmed by the MS. BULL. and
feveral in the figure are from the Parifian MSS. communicated
by Dr. MOOR. Before giving the particular folutions of
the Deliacal problem, PAPPUS premifes a ftatement of the
ancient diftinction of lines, already explained. In this
account he mentions the folution of the Deliacal problem
by means of the *Conic Sections*, without however giving a
detail of that folution; but he explains feveral others, viz.
three mechanical, by ERATOSTHENES, HERO, and PAPPUS
himfelf; and a fourth of NICOMEDES, by the defcription
of another curve, fuppofed to be invented by him, and
called the *Conchoid*. EUTOCIUS, in his commentary on
ARCHIMEDES (Prop. 2. lib. ii. *de Sphæra et Cylindro)*
gives feveral other folutions, (ten in all,) and particularly
two by MENECHMUS by the interfection of two *Conic
Sections*.—It may be obferved alfo, that in fome of the MSS.
of PAPPUS, at the end of the third book are placed fome
variations of the mechanical folutions of this problem, not
taken notice of in COMMANDINE's edition.†

† In Prop. 24. lib. iv. The folution of this problem by the conchoid is repeated;
and the folution in PAPPUS's method in this book is alfo repeated in lib. viii. Prop. 11.

The fecond problem refpecting the *medietates* is branched
out into twenty-one Propofitions, with feveral definitions
and explanations. He diftinguifhes three fpecies, and when
three lines (or three numbers) are either arithmetical,
geometrical, or harmonical proportionals, according to the
common definitions of thefe terms, they were anciently
called *medietates:* while the term *analogy* was ufually
applied only to the geometrical proportionals; fo that ftrictly
an *analogy* is a *medietas*, but not converfely.

The principal problem‡ which he propofes, is to place the
three medietates in a femicircle. The refolution of this
by PAPPUS is in Prop. 16th, but he begins with animadverting
on a geometer of his time for his manner of treating the
problem, though unjuftly, as Dr. SIMSON obferves in a note
on this paffage.§

Before this 16th Prop. however are ten problems according
to COMMANDINE's enumeration, refpecting the three medietates. But after that Propofition he obferves, that NICOMACHUS
the Pythagorean, and others, had treated not only of thefe
original medietates, but alfo of three* others which were added

‡ Called the fecond: (The Deliacal problem being the firft.) It is propofed in
fol. 7 b. COMMAND. PAPPI. And in the fame page is the definition here referred to.

·§ Dr. SIMSON quotes VINCENTIO VIVIANI *de Locis folidis*, lib. iii. page 90, where
there are fome Propofitions concerning the medietates, and alfo a remark of this
miftake of PAPPUS. I may add, that BLONDEL, in his *Resolution des 4 Problemes
d'Architecture*, makes the fame corrections of PAPPUS's cenfure; and adds, that the
folution by PAPPUS (Prop. 16.) is defective. But his objection to that Prop. is rather
verbal than geometrical. See BLONDEL, p. 36. 37.

* PROCLUS, in his *Commentary* on EUCLID, (p. 19, HERVAGII,) ftates that EUDOXUS
of Cnidus, about the time of PLATO, added three analogies (τρεις αναλογιας) to the three
then known, and the former probably were the medietates alluded to in this place
by PAPPUS.

among the ancients, and four more among the moderns, making ten altogether, which he defines and illuftrates by a number of Propofitions to the 27th inclufive. Thefe, particularly the ancient medietates, were much confidered in the Platonic School, in which probably they originated, and to which PAPPUS belonged. ERATOSTHENES compofed two books on the medietates, (probably the *Loci ad medietates*) and though they were part of the τόπος ἀναλυομένος, PAPPUS gives no defcription of them, and has preferved no Lemmata connected with them.† This may be confidered as a prefumption of his opinion that in his time they were not regarded as important ; and he further remarks of the more ancient medietates, that they were ufeful, chiefly in explaining allufions to them in ancient writings; and probably this was the principal object of the details which he has given refpecting them.

The general nature of the third problem has been mentioned, and it is expanded into fixteen Propofitions. It is a fort of paradox, and he refers to a treatife of ERYCEMUS on Paradoxes, which, in the time of PAPPUS, was generally circulated, and probably in fome eftimation. Some of the Propofitions are curious, but in the prefent ftate of the fcience they cannot be confidered important, and are interefting only as examples of the geometrical fpeculations of former times, (Prop 28—43.)

---

† From the obfervation of PAPPUS at the end of his defcription of the *Inclinations* of APOLLONIUS, it would feem that thefe *Loci ad medietates* were *plane*, and, in the order of the τόπος ἀναλυομένος, placed laft.

The fourth and laft general problem of this book is to infcribe in a fphere the five regular folids; and as thefe folids were objects of much fpeculation in the Platonic fchool, they were naturally attended to by PAPPUS, who belonged to it. This problem occupies the remainder of this book in the laft fixteen Propofitions, of which eleven are preliminary to the other five; and thefe laft contain the folution of the problem refpecting the before-mentioned five regular folids.

§. III. *Of the Fourth Book.*

THE fourth book is mifcellaneous, containing theorems of the three ancient claffes of Propofitions formerly mentioned; plane, folid, and linear.* The firft may be remarked as an extenfion of the 47. 1 Elem.; and the 4th is a curious theorem, with an analyfis and compofition, which was rendered more general by Dr. M. STEWART, in a paper publifhed in the firft volume of the Edinburgh *Phyfical Effays*;§ and fome Propofitions alfo are added by Dr. SIMSON, in his notes.

* The title in MS. BULL. is as follows: Πασων συναγωγης δ' οπιρ ιςιν αθηρα‡ θηρη- " μαιων ιπιπεδω, και ςιρεων και γραμμικων."

‡ Probably it ought to be ανθηραν, as it is in the title of the eighth book.

§ In Dr. STEWART's Paper, in that volume, are feveral Propofitions connected with this propofition of PAPPUS, and which are analyfed and demonftrated in the ancient manner. A fcholium, containing fome conic theorems without demonftrations, is added; and at the end of it is a theorem, alfo without demonftration, which is truly a Porifm, and is analyfed and demonftrated by Dr. SIMSON, Prop. 91 of his treatife. Dr. STEWART, from delicacy no doubt to Dr. SIMSON, who wifhed to keep his expo-

U

The 10th is the cafe of one of the problems of the *Tactions* of APOLLONIUS, in which three circles touching each other are given, to which the three preceding Propofitions are preliminary; all however requiring the corrections remarked by Dr. SIMSON.†

Before the 13th Prop. a property of the figure called *Arbelon*‡ is ftated, as an ancient Propofition generally known, and of which the demonftration is in the 16th, the intermediate Propofitions being neceffary to it.

At the 19th Prop. he gives the definition of the Spiral of CONON, and demonftrates fome of its principal properties.‖ This is the curve fully treated of by ARCHIMEDES, and in a fubfequent part of the book it is employed in the folution of fome problems. PAPPUS (Prop. 24) repeats the definition of the

'fition of the Porifms fecret till he fhould publifh his treatife, ftates the Propofition as a theorem, though he was well acquainted with Dr. SIMSON's difcovery, and had investigated many curious Porifms.——See *Edinburgh Phyfical Essays*, vol. i. p. 141—172.

† There are fome amendments on the 12th, neceffary to make the demonftration complete.

‡ The Arbelon (ἀρβῶος) is mentioned in the *Lemmata* of ARCHIMEDES, (Prop. 4, 5, 6;) and it is remarkable, that though PAPPUS in his *Collections*, and particularly in this book, often quotes ARCHIMEDES, there is no allufion to the *Lemmata* in his long and curious difcuffion of the properties of the Arbelon. There is alfo a lemma of COMMANDINE's, for demonftrating a Propofition of PAPPUS on this fubject, (Prop. 14,) which is the firft Propofition of the *Lemmata* of ARCHIMEDES, and which COMMANDINE afterwards (fol. 52, b.) afferts to be compofed by himfelf. From this it may be inferred, that COMMANDINE certainly had never feen the *Lemmata* of ARCHIMEDES, and moft probably neither had PAPPUS. The *Lemmata* have never been found in Greek, and have by feveral learned men been fuppofed not to be the work of ARCHIMEDES; and the circumftance now mentioned favours that fuppofition.

‖ In Prop. 21, refpecting the Spiral, the demonftration is conducted on principles like the method of Indivifibles of CAVALLERIUS.

conchoid of Nicomedes, and the solution of the Deliacal
Problem by it, as in the preceding book: he afterwards em-
ploys this curve for refolving another celebrated problem of
antiquity, the trifection of an arch of a circle, which was found
impracticable by plane geometry; and which, with the Deliacal
Problem, roufed the efforts of mathematicians to inveftigate
new curves for refolving them.

He proceeds to explain the origin and properties of the
quadratrix, (τὴ ραγωνίζετα,) affumed by Dinostratus and
Nicomedes, fubfequent geometers, for the quadrature of the
circle, from which it obtained its name. But he obferves
that another geometer, Sporus, was not fatisfied with this
application of that curve, and he gives fome detail of the objections propofed by Sporus; and particularly, that in the
conftruction of this line the problem to be refolved by it is in
fome meafure affumed. Pappus however, befides the common
defcription of its origin and of its properties, gives what
he confiders as a more ftrictly geometrical defcription, by
*Loci ad Superficiem.**

After feveral Propofitions on this fubject, there is a repetition
of the ancient claffification of lines, and nearly in the fame
words† in which it was ftated in the third book. It feems to
be introduced here for illuftrating the problem of the trifection
of an arch of a circle, as in the former book it had a more
particular reference to the Deliacal problem.

* This is Prop. 28, which is fo much corrupted in Commandine as to be fcarcely
intelligible; and in Note G. at the end of the preceding Memoir is placed Dr.
Simson's corrected edition of it. Another defcription of it is given in Prop. 29.
† The Greek of this paffage is added in Appendix II.

This problem is refolved by PAPPUS in various ways; firft, by a folid inclination, to which Dr. SIMSON adds an improvement; then by an hyperbola without the inclination;* and he mentions that in another treatife† he had refolved it by means of the conchoid, of which the method is obvious.

He fubjoins fome linear problems, fuch asdividing an angle or arch of a circle in a given ratio ; the conftruction of an ifofceles triangle, of which each of the angles at the bafe fhall have a given ratio to the angle at the vertex ; which are eafily refolved, by affuming the quadratrix or fpiral, as proper means of geometrical folutions.

Befides the very important emendation of Prop. 28th and 29th of this book by Dr. SIMSON, there are various other neceffary corrections in other Propofitions, and fome ufeful readings adopted by him from MS. BULL. which it is unneceffary to fpecify. At the end of this book, there is in MS. BULL. a very imperfect fketch of a problem faid to have been ufed by ARCHIMEDES, in his attempt to find a ftraight line equal to the periphery of a circle; and alfo that this ufe of it had been animadverted on. It is not in COMMANDINE's tranflation, and in its prefent ftate is altogether unintelligible.§

---

* This Inclination is alluded to in Note F. p. 99.

† From the very brief notice of this work, it feems to have been a comment on the Analemma of DIODORUS: " και ημεις εν ω εις τα αναλημμα Διοδορε τριχα τεμειν την " γωνιαν βυλομενοι κεχρημεθα τη προειρημενη γραμμη." MS. BULL. Vide COMMAND. PAPP. fol. 56, a. 1. 6.

§ This Propofition, in the fame unintelligible ftate, is found in Savil. MS. No. 3, and at the end of the fourth book.

§. IV. *The Fifth Book.*

To the fifth Book there is a preface, in which PAPPUS makes fome philofophical obfervations on the curious inftincts of animals, which in many cafes fupply the place of reafon. He mentions more particularly the inftinct of bees, by which they conftruct the receptacle for their provifion on geometrical principles, by employing the hexagon, which of the three equilateral and equiangular figures that occupy the fpace round a point, fupplies, with the fmalleft labour, the moft convenient accommodation. This obfervation feems to be introductory to the purpofe of the book, which, at the end of the preface, he ftates to be; to prove that of plane figures which are equilateral and equiangular, and have equal perimeters, the greater fpace is contained by the figure with the greater number of fides; and that of all plane figures, of equal perimeters, the circle is the greateft.*

This fubject of ifoperimetrical figures is largely treated of in this book, containing fifty-feven Propofitions.‡ Befides

* The title of the fifth book in MS. BULL. is the following, Πάππου 'Αλ ξανδεῖς συναγωγῆς ἕ, περιέχει δὶ συγκρισᾶς τῶν ἴσων περ'μᾶρον ἐχουῖῶν ἐπιπεῖδῶν σχήμάῖαι πρὸς ἀλλήλα τι καὶ :ὸν κύκλον,‖ καὶ συγκρισᾶς τῶν ἴσων ἐπιφανιαι ἐχουῖῶν ῖερῶν σχήμάῖων πρὸς ἀλλήλα καὶ τῆν σφαῖραν.

‖ Note in MS. BULL. it is erroneonsly. τῶν κυκλῶν, but in the Savil. MS. No. 3. it is τὸν κύκλον.

‡ In Prop. 1. an eafy Propofition is affumed, which is demonftrated in a Lemma (Note B. on that Prop.) by COMMANDINE. It is remarkable, that a Propofition of the fame import in Greek is found on the margin of MS. BULL. And in the text of MS. SAVIL. No. 3, is a fimilar Propofition. It would require however the examination of other MSS. to afcertain the hiftory of this Lemma.

the object mentioned in the preface, several other things are
confidered. It is proved, that of plane figures with equal
perimeters, the greateft is that which is equilateral and
equiangular. This principle is extended to folids; the
regular folids (as they are called) are compared, and it is
proved, that, of thofe with equal furfaces, the greateft is that
with the greater number of fides. Thefe Propofitions are
meant no doubt to be introductory to a demonftration of the
Propofition, that of all folids with equal furfaces the greateft
is the fphere. There are many references to ARCHIMEDES,
and feveral of the Propofitions in his treatife of the Sphere
and Cylinder are demonftrated in this book by PAPPUS,* and,
as he mentions, in a different manner.

It was formerly obferved, that a number of Propofitions in
it, refpecting the comparifon of the five regular bodies having
equal furfaces, were demonftrated only fynthetically from
the brevity and the facility of communication in that method;
and that this explains the omiffion of the analyfis in many
ancient Propofitions, though there can be no doubt of that
analyfis having been ufed by the authors in the inveftigation
of them. He introduces the Propofitions refpecting thefe
regular folids by a general differtation on them before Prop. 18.
in which he profeffes great reverence for the doctrines of the
divine PLATO, in whofe fchool it is fuppofed that thefe bodies
were firft defined, and their relations fully treated of. He

* In Prop. 35. (fol. 96. COMMAND.) is a very material improvement of the demonftration by Dr. SIMSON, by which the long note (Z) of COMMANDINE becomes
unneceffary. This may be remarked in many of the Doctor's corrections, which,
even when fhort, often fuperfede the ufe of long and fometimes tedious comments
by COMMANDINE.

APPENDIX I. 151

remarks, that it is affumed by philofophers, without proof, that
the fphere is the greateft of folids, having equal fuperficies: he
proceeds to give fome popular illuftrations of that truth; and
in Prop. 18. he propofes a comparifon of the fphere with the
regular folids. In the end of the book he ftates, what is eafily
afcertained, that there can be only five regular folids, which
are comprehended by equal, fimilar, and equilateral polygons,
viz. the tetraedron, cube, octaedron, dodecaedron, and icofa-
edron; in fome of the Propofitions Dr. SIMSON has remarked
feveral proper and neceflary corrections.

§. V. *Sixth Book.*

THE fixth Book of PAPPUS, as the title intimates,‡ is employed chiefly in explaining and correcting fome Propofitions
of THEODOSIUS, and fome other ancient writers, in treatifes
containing chiefly, what is popularly called, the doctrine of
the fphere. In a fhort preface the object of this book is
ftated, with a reference to three examples of Propofitions
criticifed in it, viz. the 6th, iii. THEODOSIUS *on the Sphere*;
the fecond Prop. of EUCLID's *Phænomena*; and the fourth
Prop. of THEODOSIUS *on Days and Nights*. This collection
of treatifes, obtained, in the Alexandrian fchool, the name of
μικρὸς ἀςρονόμος, or, as it is in PAPPUS μικρὸς ἀςρινομύμενος. Among
the Arabians they were called *Libri intermedii inter* ςοιχοιωΐὴν *et
magnum conftructorem*, that is, between EUCLID and PTOLEMY:.

‡ The title of this book in MS. BULL. is Πάππου 'Αλεξανδρεῶς συναγωγῆς ἑκτον, περιέχιι
ἧ ἦν ἐν τὰ μικρὰ ἀςρτοιμεμένινω θιωριμαύων ἀπορῶν λύσεις.

and in contraſt to the greater work of the latter (the Almageſt, ſometimes called μέγας ἀςρονόμος;) theſe ſmaller tracts got the name of μικρὸς ἀςρονέμος.

According to Vossius, this collection contained the nine following works, viz. THEODOSII *Sphærica*, EUCLIDIS *Optica*, ejuſdem *Phænomena*, THEODOSII *de Habitationibus*, ejuſdem *de Noctibus et Diebus*, AUTOLYCUS *de Sphæra*, ejuſdem *de Ortu et Occaſu*, ARISTARCHI *de magnitudinibus et distantiis Solis et Lunæ*, HYPSICLIS ἀναφορικὲν, ſive *de ascensionibus*.† FABRICIUS, when ſtating this collection, includes in it alſo MENELAI *Sphærica*, and EUCLIDIS *Data et Catoptrica* :‡ and he adds, that theſe treatiſes, or moſt of them, are often found together in Greek MS. in the libraries of Italy and France.

It may be alſo obſerved, that all theſe treatiſes, except that of HYPSICLES, are quoted or alluded to in this ſixth book, in which PAPPUS examines ſeveral ἔνςασις *(Instantiæ)*, to be found in them. By this term he appears to underſtand the objections to certain Propoſitions in theſe writers, or the limitations and exceptions neceſſary to be made reſpecting them, but which had been omitted by the reſpective authors. PROCLUS gives an explanation of this term;§ and from his diffuſe and not very ſatisfactory account, this word ἔνςασις *(Instantia)* ſeems to have been applied to any objection made to a Propoſition, or even to the denial of it, as preparatory to an indirect demonſtration of it, by proving the abſurdity of

† GER. VOSSIUS *de Scientiis Mathematicis, et Chronologia Mathematicorum*, çap. xxxiii. §. 18.

‡ FABRICII *Biblioth. Græc.* Harles, Hamb. 1795. tom. iv. p. 16, and alſo p. 212. SAVILIUS in Prælect. II. in EUCLIDEM.

§ PROCLUS, p. 58. HERVAGII; p. 121. BAROCII.

that denial. In PAPPUS, however, the application feems to be confined to the former meaning, with which the examples which he gives, from THEODOSIUS and others, correfpond.†

Prefixed to Prop. 39 of this book, is a long difcuffion concerning the pofitions in the before mentioned tract of ARISTARCHUS SAMIUS and this, with the 39th, 40th, and 41ft Propofitions on the fame fubject, is inferted by Dr. WALLIS in his edition of *Aristarchus* publifhed at Oxford in 1688. He adds the Greek of this extract from PAPPUS according to the two Savilian manufcripts, which he takes an opportunity of characterizing.*

In the explanation and correction of feveral paffages in thefe tracts by PAPPUS are introduced fome curious Propofitions, by which the doctrine of the fphere of that period was improved; and there are fome Propofitions relating to a fort of perfpective or projection of the fphere, which are connected with Propofitions of EUCLID's *Phænomena.* It may be alfo obferved, that in four Propofitions, viz. 31, 32, 33, 34, which are preliminary to fome following Propofitions refpecting the fphere, fome diftinctions of magnitudes are ftated as examples, which may be either increafed or diminifhed without limit; or may be increafed indefinitely,

† Examples of *instantiæ* may be found in Prop. 21, of this fixth book, to which the preceding ten Propofitions refer. Alfo in Prop. 51, (fol. 145. a. at the bottom, COMMAND.) Other criticifms on EUCLID's *Phænomena*, and alfo on HIPPARCHUS, are mentioned in fubfequent Propofitions of this book. There are alfo many references to PTOLEMY.

* In his *præfatio ad Lectorem* he obferves, " quorum (fc. MSStorum) qui " elegantius fcribitur (fc. No. 3.) eft mendofior ; quique feftinantius et minus " eleganter fcribitur (fcil. No. 9.) eft emendatior."

X

while there is a limit to the decreafe, and converfely. And I mention them here, merely to remark the inaccuracy of COMMANDINE's tranflation, by introducing the term *infinite* without authority from the original. Dr. SIMSON obferves that he had not confidered this book fo particularly as he purpofed afterwards to do, but which he appears not to have accomplifhed : at the fame time he had made fome neceffary and ufeful corrections in feveral Propofitions as they ftand in COMMANDINE's tranflation.*

§. VI. *Seventh Book.*

THIS Book, as the title imports, is wholly employed on that curious fubject, the ancient analyfis.†

THE collection of treatifes known by the name of *τόπος ἀναλυομένος*, next to the *Elements* and *Data* of EUCLID, were important for facilitating the refolution of geometrical problems. They were compofed, by the elder ARISTÆUS, EUCLID, and APOLLONIUS;‡ but at what time the collection

* It may be obferved, that in the MS. BULL. and in the Savil. MS. No. 3, are fome marginal notices and at the end of Prop. 59, a fcheme refpecting climates which may deferve confideration; in that Propofition alfo are references to PTOLEMY.

† The title of the 7th book, from one of the Parifian MSS. No. 2368, is as follows; Πάππου Ἀλεξανδρέως συναγωγῆς ζ ὃ περίχει τὴν τάξιν, καὶ τὴν περιοχήν καὶ τα λήμματα τῦ ἀναλυομίνου τόπου.

‡ In this firft ftatement of the authors in the preface to this viith book thefe three only are mentioned. In a fubfequent more particular account (alfo in this preface) PAPPUS mentions two books on the *Medietates* by ERATOSTHENES. But neither this work of ERATOSTHENES, nor the *Loci* of ARISTÆUS, nor the *Loci ad Superficiem* by EUCLID, are defcribed by PAPPUS.

obtained the name by which they are diftinguifhed in the preface to this book, is unknown. They muft have been compofed at different periods, from the known interval of the times of the refpective authors, who all lived above 500 years before the age of PAPPUS. The expreffion of PAPPUS is not definite; yet it feems to imply that the collection was made, and the name impofed, before his own time, for his expreffion feems to fuppofe the order of the books as exifting.‡

The preface to this book, one of the moft valuable remnants of ancient geometry, contains a particular defcription of the nature and contents of a certain number of thefe treatifes, which, both from the fubjects of them, and from their being thus diftinguifhed by PAPPUS, may be confidered as by far the moft important part of the collection; and the book itfelf contains a number of Lemmata or elementary Propofitions affumed (probably without proof) in the treatifes defcribed in the preface, but of which the demonftrations are added by PAPPUS.

PAPPUS, however, premifes a fhort but accurate account of the ancient method of analyfis and fynthefis; of which a free tranflation will be more fatisfactory than any abridgement, for conveying a correct notion of this curious branch of fcience, fo much valued among the ancients, and, till the

‡ Dr. HALLEY, whofe authority in fuch points is very great, at the end of his preface to the *Sectio Rationis,* intimates an opinion that the collection was made by PAPPUS. But the circumftances now mentioned render this inference from the words of PAPPUS at leaft doubtful; and Dr. SIMSON has remarked another miftake of Dr. HALLEY's in the before mentioned preface refpecting the principal object in forming this collection.

x 2

time of Dr. HALLEY and Dr. SIMSON, fo little underſtood
among the moderns.

It is employed both for the refolution of problems, and alfo
for inveſtigating the truth of theorems either aſſerted or
conjectured to be true; though it is in the former claſs of
Propofitions that the ufe and importance of this method is
chiefly known.

In the analyfis of a problem, what is propofed to be done,
or to be found, is fuppofed to be obtained; and the confequences of fuch an aſſumption are fucceſſively deduced by
means of all previouſly known practical Propofitions, which
appear to be connected with it, till at length we arrive at
fome conclufion, which, from the ſtate of geometrical fcience
at the time, we know to be practicable; and thence the
folution of the problem is deduced Synthefis, or compofition
is the reverfe of analyfis; and by aſſuming the laſt practicable
confequence of the analyfis, we proceed in a contrary order
through the feveral ſteps of that analyfis till we neceſſarily
reach the conſtruction of the thing required, and then the
problem is refolved. In this analyfis however, if we arrive
at a confequence which we know, from previous Propofitions,
to be impoffible, then the problem propofed muſt itfelf be
impoffible; and further, if we find, from the progrefs of the
compofition, that in certain relations of the given magnitudes
the conſtruction is practicable, while in others it becomes
impoffible; the afcertainment of thefe relations becomes a
neceſſary part of the folution, and is called the determination
of the problem.

In the analyfis of a theorem,* the affertion in the enunciation is affumed as true; and by reafoning from it, by the application of known geometrical Propofitions which appear to be connected with it, we trace the fucceffive confequences of the firft affumption, till we arrive at fome one which we know to be true, or to be falfe. If we arrive in this manner at a true Propofition, by proceeding from it, in a contrary order, through the feveral fteps of the analyfis, we fhall neceffarily arrive at the propofed affertion; and this laft proceeding will be a demonftration or compofition of the theorem. In like manner, if in this inveftigation we arrive at a conclufion which we know to be falfe, from the fame neceffary concatenation of Propofitions, we infer the falfehood of the affumed affertion; and if it were requifite, we might alfo demonftrate that inference. But it not being my defign, in this general account of the feventh book of PAPPUS, to attempt a full expofition of the ancient analyfis, which, for the ufe of thofe who have not confidered the fubject, would require alfo the illuftration of examples, I can only refer to thofe treatifes where fuch explanatory examples of the practice of analyfis may be found.†   I fhall

* The analyfis of theorems is often ufeful. A theorem may be propofed to a geometer, that he may inveftigate the demonftration; it may be only fufpected to be true from analogy, from the apparent relations of magnitudes in geometrical diagrams, and even from accident; but in all fuch cafes an analyfis will afcertain the truth or falfehood of the affumed Propofition.

† In the remaining works of APOLLONIUS, as publifhed by Dr. HALLEY, and alfo in PAPPUS, many problems and theorems are treated analytically. In Dr. SIMSON's different works are various examples in the pure ftile of the ancient geometry. The analyfis of theorems is well illuftrated in Dr. STEWART's *Propofitiones Geometricæ more veterum demonftratæ*. For problems, I may refer to Dr. HORSLEY's (Bifhop of St. Afaph) *Delectus Problematum*; and to Mr. Profeffor LESLIE's *Geometrical Analyfis* in his *Elements of Geometry*. In that treatife the general problems of APOLLONIUS, mentioned in this book of PAPPUS, are introduced as examples.

therefore only further obferve, that, in this practice, fome of
the principles and rules commonly laid down by writers on the
application of the algebraical analyfis to the folution of geometrical problems, may here alfo be ufefully employed;
though feveral of them, no doubt, are more peculiarly fitted
for the modern fyftem.†

After this general account of the ancient analyfis, PAPPUS,
in his Preface, proceeds to enumerate the books contained in
the τόπος ἀναλυομένος, and afterwards to give a particular
account of the nature and contents of a certain number of
them, viz. of twenty-four; the whole number being thirty-
three.‡

Thefe twenty-four books are, I. *Data* of EUCLID; III
books *De Porifmatis*, alfo by EUCLID; and the following
twenty by APOLLONIUS: II. *De Sectione Rationis*; II. *De
Sectione Spatii*; II. *De Sectione Determinata*; II. *De Tactionibus*; II. *De Inclinationibus*; II. *De Locis Planis*; and VIII.
De Conicis.

Thefe books, though all ufeful in facilitating the folution of
geometrical problems, yet are of different characters, and pro-
mote the object of the whole collection in different ways; as
may appear from the many references to them in this Memoir,
of which the following fhort recapitulation is added.

† See NEWT. *Arith. Universalis*, cap. 23.——See alfo fome obfervations of Dr.
SIMSON on the ancient analyfis, in Note K. at the end of the Memoir, p. 121—128.

‡ In this preface PAPPUS mentions his purpofe of giving a defcription of thefe books
as far as the *Conics* of APOLLONIUS; and therefore, according to his arrangement,
which is not in the order of time, he appears to omit by defign the nine following
books: V. of ARISTÆUS *de Locis Solidis*, (which were no doubt fuperfeded by the
Conics of APOLLONIUS;) II. of ERATOSTHENES *de Medietatibus*; and II. by EUCLID,
de Locis ad Superficiem.

I. *The Data of* EUCLID. This book, one of the few which have been preferved in the original language, may be confidered as a collection of elementary problems; and the demonftrations as given by EUCLID, would become the analyfes of thefe Propofitions, had they been enunciated as problems. The book was conftantly ufed by the ancients in their refolutions of problems, though it was not their practice to quote the particular Propofitions; and this application of the Data was continued among fuch of the moderns as follow ftrictly the ancient method of analyfis, and is now fo well known, that it is unneceffary to give any detail of it.* The *Sectio Rationis*, *Sectio Spatii*, *Sectio Determinata*, and the Treatifes *De Tactionibus*, and *De Inclinationibus*, are all general problems of frequent recurrence in geometrical inveftigations, and were refolved by APOLLONIUS in the moft complete manner, all the poffible cafes being diftinguifhed; and of each cafe a feparate analyfis and compofition are given, with the refpective determinations in all thofe cafes which required them.†

The ufe of thefe general problems, as has been repeatedly mentioned, was for the more immediate refolution of any propofed geometrical problems which could be eafily reduced to a particular cafe of any one of them.‡ By fuch a

---

* See in Note K. at the end of the Memoir, fome remarks on the Data, in the correfpondence between Dr. SIMSON and Mr. SCOTT.

† PAPPUS, in this preface, alluding to a problem connected with the Treatife of Tactions, ftates with brevity, but with precifion, what is requifite in a perfect folution of fuch general problems: " καὶ ταῦτα ἀναλῦσαι, καὶ συνθεῖναι, καὶ διορίζεσθαι καλᾶ στᾶσιν."

‡ When a problem can be refolved with equal facility by the ufe of known elementary Propofitions, a reference in fuch a cafe to any of thefe general problems becomes unneceffary.

reduction the propofed problem was confidered as fully refolved;
becaufe it was then neceffary only to apply the analyfis, compofition, and determination of that cafe of the general problem
to this particular Propofition, which was fhewn to be comprehended in it. The apparatus of feparate folutions, with the
determinations of every poffible cafe which is effential to the
ufe of thefe general problems, may appear forbidding; and if
regularly perufed without examples of the application, may
fometimes appear tedious and uninterefting; and this perhaps,
may have created fome prejudice againft the ftudy of them.
Dr. HALLEY, indeed, feems to think that the books of the
τόπος ἀναλυομένος, were intended merely for the inftruction of
beginners in the ftudy of the geometrical analyfis:§ but it is
juftly obferved by Dr. SIMSON, that though they may be moft
advantageoufly employed for that purpofe, yet that it is mani-
feft from the contents of thofe books, of which he gives fome
detail, and alfo from the manner in which they are referred to
in the fmall remains which we poffefs of the ancient geometry,
the chief object of them was what has now been ftated. Even
in PAPPUS there are fome examples of problems being refolved
by a reduction of them to cafes of thefe general problems. In
the 85th Prop. of this feventh book, a problem, ufeful in the
Treatife of Inclinations, is by analyfis brought to a cafe of the
Sectio Determinata, as was formerly mentioned in the Memoir.
Another example is Prop. 164, of the fame book, the laft lemma
belonging to the Porifms; which is a problem refolved by
PAPPUS, by reducing it to a cafe of the *Sectio Spatii,* and the
compofition of the problem is confidered as compleated, merely

§ HALLEII *Sect. Rationis,* Præfat. See alfo *Loci Plani,* SIMSON's Præfat. p. vij.

by a reference to the conftruction of that cafe. This tract of
APOLLONIUS being loft, and no reftoration of it having been made
in the time of COMMANDINE, the cafe referred to is refolved by
him, though in a tedious manner, which may be compared with
the folution of it in Dr. HALLEY's reftoration of that work.*

The other books, defcribed in this preface by PAPPUS, are
ufeful for the fame purpofe of facilitating the folution of geometrical problems, though in a manner different from what
has juft been mentioned of the preceding treatifes. The treatifes of *Loci* in particular are very important, and in a method
well underftood by thofe even flightly acquainted with either
the ancient or modern geometry. The *Loci Plani* of APOLLO-
NIUS are ufeful for the folution of plane problems; and often
alfo, with *Loci* of a higher order, may be required in the
folution of fuperior problems. Though the original work of
APOLLONIUS be loft, the elegant and compleat reftoration of it
by Dr. SIMSON leaves nothing to be regretted on that fubject.
The *Loci Solidi* of ARISTÆUS might have been very ferviceable
in the folution of folid problems, but it has perifhed; and
probably, even in the time of PAPPUS, had been fuperfeded by
the great work of APOLLONIUS on *Conic Sections*, from which
not lefs important aid might have been derived for refolving
that clafs of problems.†

* SCHOOTEN alfo refolves this problem, *Exercit. Mathem.* p. 104, 105, and feems
to blame PAPPUS for not refolving it directly, but by a reference to a cafe of the *Sectio
Spatii*; from which it appears, that SCHOOTEN was not apprifed of the true ufe of
thefe ancient books. His Propofition alfo is not fo general as it is ftated in PAPPUS.
See Dr. SIMSON's edition of this Prop. *Opera Reliqua*, p. 529.

† As feven books of the *Conics* of APOLLONIUS have been recovered, and the
eighth reftored in a fatisfactory manner by Dr. HALLEY, it is unneceffary to enter

The porifms of Euclid were a peculiar clafs of Propofitions, and ufed among the ancients in the folution of fome of the moft difficult geometrical problems. The long account of them in this preface has been particularly unfortunate in fuffering from the injuries of time; fo that till Dr. Simson's perfevering induftry and ingenuity were employed on them, no fatisfactory explanation of the nature of thefe Propofitions could be difcovered, nor could any fingle Propofition of Euclid's treatife be reftored. But in the preceding Memoir fo full an account of Dr. Simson's reftoration of them has been given, that it is unneceffary in this place to make any further obfervations refpecting them.

With refpect to the τόπος ἀναλυομένος, I fhall only further obferve, that on the revival of mathematical learning in Europe, if the ftudy and application of the ancient analyfis had continued, without being nearly fuperfeded by the ufe of the modern algebra; it is highly probable that later geometricians would, from time to time, have made additions to the ancient collection; and would have inveftigated various other general problems, with compleat folutions of their feveral cafes, for the fame general purpofe for which thefe ancient books were compofed by Euclid and Apollonius.

At the end of the defcription of the *Conics* of Apollonius, in this preface to the feventh book of Pappus, there are fome interefting obfervations of a more general nature, which have

into any difcuffion refpecting the account of this work by Pappus. It is, however, both curious and interefting; and Dr. Halley, in his preface to the corrected edition of this preface to the feventh book of Pappus, makes fome valuable obfervations on it. Dr. Simson adds fome ufeful notes on the *Lemmata* of this work.

been particularly remarked by Dr. HALLEY. I allude to the account of the celebrated ancient problem of the " *Locus ad* " *tres et quatuor rectas*;" and the extenfion of it mentioned by PAPPUS to the cafes where there are a greater number of ftraight lines than four, to which lines are drawn in given angles from the point, of which the *Locus* is to be inveftigated. The original *Locus* is folid, but thefe other cafes produce *linear Loci*, that is, curves of fuperior orders; which, however, had not been inveftigated in the time of PAPPUS. In fome general obfervations on the fubject by PAPPUS, it may be remarked that he mentions the method of expreffing dimenfions above the cubic, by means of compound ratios.

This laft article of the preface is concluded by a diftinct enunciation of the celebrated theorem of GULDINUS, which, twelve centuries after the age of PAPPUS, excited much curiofity and admiration. Whether this theorem was invented by PAPPUS, or by fome other geometer, is not ftated. It is proper to remark, however, that no imputation is conveyed on the originality of the difcovery by GULDINUS; as in COMMAN-DINE's tranflation, which was the only account then in print, the Propofition is altogether unintelligible.* But as this portion of the preface has not been confidered in Dr. SIMSON's notes,

* The improved tranflation of this preface of PAPPUS, by Dr. HALLEY, appeared only in 1706, long after the time of GULDINUS; and in it this theorem from PAPPUS was firft intelligibly publifhed. MONTUCLA indeed remarks, (tom. i. p. 325,) that the work of GULDINUS was publifhed before the fecond edition of COMMANDINE's PAPPUS in 1660. But that edition could not have given him any aid, for it contained no valuable improvement of the former; and in this particular point, it is an exact tranfcript of the paffage in the firft edition.

it becomes unnecefſary in this place to enter into any further difcuſſion of it.

Having given ſo particular an account of the preface of the ſeventh book, (certainly one of the moſt curious remains of ancient geometry,) a very ſhort ſtatement of the contents of the book itſelf will be ſufficient. It conſiſts of a great number of Propoſitions, (238,) ſome of which are problems, but the greater part theorems. They are called *Lemmata*, and divided into claſſes according to the ſeveral treatiſes deſcribed in the preface, to which they reſpectively belong. They appear in general to be elementary Propoſitions aſſumed, but we may ſuppoſe not demonſtrated, in theſe ancient books, from which they were collected probably by PAPPUS; and the collector has added the demonſtrations. Some of them indeed are curious and important Propoſitions; but the greater number of them are chiefly valuable for the aid which they give in the attempts of modern geometers in reſtoring accurately the loſt books of the τίπος αναλυομενος. In ſuch attempts, it was an intereſting object, not merely to reſolve the general problems contained in that collection, with their caſes and determinations, but alſo, where it was poſſible, to follow the particular mode of ſolution employed by the original authors. Theſe Lemmata‡ preſerved by PAPPUS have

‡ According to PROCLUS, in his comment on Prop. 1. I. Elem. a Lemma (ſump-tio BAROC.) is an aſſumption of a geometrical truth, in the demonſtration of a Propoſition; which truth, however, has not been demonſtrated by known geo-metrical writers. In this ſenſe is the term uſed by PAPPUS, and the Lemmata of the ſeventh book ſeem to be of that deſcription. It appears from the books of APOLLONIUS's *Conics* ſtill preſerved, that the Lemmata of PAPPUS belonging to

fupplied the means of doing this; and when proper analyfes and compofitions were difcovered, which required alfo the -ufe of the particular Lemmata mentioned by PAPPUS, there was a high degree of probability, that fuch folutions were truly thofe given by EUCLID and APOLLONIUS.

Many neceffary corrections of thefe Lemmata, as they appeared in COMMANDINE's tranflation, were made by Dr. SIMSON.   Some of the corrections were fupplied by the Parifian MS. of PAPPUS, of which Dr. MOOR procured a copy, as has formerly been mentioned; and they are remarked in the Doctor's reftorations of the *Loci Plani*, and *Sectio Determinata*. In the Lemmata belonging to the Porifms, which are all given in his pofthumous work on that fubject, and in an improved ftate, are many important emendations; but it does not appear

that work were affumed by him, in the manner now mentioned; and Dr. HALLEY, in his edition of APOLLONIUS, gives thefe Lemmata from the Savilian MSS. of PAPPUS. In the demonftrations of the Propofitions of APOLLONIUS, Dr. HALLEY refers occafionally to thofe Lemmata, as the want of them appeared in the text for rendering the demonftrations compleat. It appears likewife from the fame edition, (p. 97 of the tranflation of vth, vith, and viith, books) that ABDOLMELEC SCHIRAZITA, an Arabian epitomizer of the *Conics*, had collected fome other Lemmata, (8.) befides thofe in PAPPUS, which were affumed without proof in the demonftrations of the feventh book, and which therefore he prefixes to that feventh book; and they are placed by Dr. HALLEY immediately after the Lemmata of PAPPUS belonging to it.

In modern times, however, a Lemma is underftood to be an eafy preliminary Propofition, which might have been incorporated with the more important one to which it is prefixed, but is more conveniently detached from it, and fometimes is itfelf a Propofition deferving notice. In the treatife of ARCHIMEDES on the Sphere and Cylinder (before Prop. 17. b i.) is an eafy Lemma demonftrated by himfelf; and at the end of Prop. 17th are feveral Lemmata, which he ftates to have been de-monftrated by thofe who preceded him; and are Propofitions of the xiith b. of EUCLID. They are in fubfequent Propofitions affumed as known, but a formal quotation of preceding works was not the practice of the ancient geometers.

that any of thefe laft were derived from that MS. as he never
quotes it ; and moft of them, indeed, were made before Dr.
Moor obtained that copy of the feventh book.‡   A confiderable number of the Lemmata of the feventh book (about 90)
have been publifhed, as corrected by Dr. Simson; and the
attention and accuracy with which he has examined thefe
Propofitions, will be obvious to the intelligent reader, by
comparing them with the edition of them in Commandine's
tranflation.   And though the remaining Lemmata in this
book do not appear to have been fo particularly confidered
by the Doctor, yet there are in his notes on them many
neceffary corrections, with ufeful explanations, fome of which
have been mentioned in the notes added to this memoir.

The Lemmata in this feventh book belong to the feveral
treatifes defcribed in the preface ; according to the annexed
arrangement from Commandine's tranflation ; in which
enumeration, however, are fome irregularities, as is obferved
by Dr. Simson.

‡ In Dr. Simson's Pappus is a memorandum for enquiry if the copy of the *Codex
Regius* contained any of the neceffary emendations which he had made on a Lemma
(Prop. 130.) of the Porifms, but no mention is made of the refult; which is a further
prefumption that the Doctor, though he had had the ufe of the MS. for correcting
the Lemmata of the two treatifes above mentioned, yet that he had not had an
opportunity of confulting it refpecting the Porifms.

Treatises.					No. of Lemmata.
No. 1.	APOLLONIUS,	*De Sectione Rationis et Spatii,* from beginning to Prop. 21 incl.			21
No. 2.	————	*De Sectione Determinata,*	to Prop.	64.	43
No. 3.	————	*De Inclinationibus,*	to —	95.	31
No. 4.	————	*De Tactionibus,*	to —	118.	23
No. 5.	————	*De Locis Planis,*	to —	126.	8
No. 6.	EUCLIDES,	*De Porismatis,*	to —	164.	38
No. 7.	APOLLONIUS,	*De Conicis,*	to —	234.	70
No. 8.	EUCLIDES,	*De Locis ad Superficiem,*	to —	238.	4*
					238

Of thefe Lemmata, fince COMMANDINE's tranflation, have been publifhed:

By Dr. HALLEY, Greek and Latin, No. 1, and 7.	91
By Dr. SIMSON, in Latin, and corrected, Nos. 2, 5, and 6.	89
By CAMERER, Greek and Latin, No. 4.	23
	203

A few others have been mentioned in this Memoir as corrected by Dr. SIMSON and others.

It is proper alfo to mention in this place, that a Propofition of fome curiofity, though not in COMMANDINE's tranflation, is found at the end of this book in the two Parifian MSS. and alfo in thofe belonging to the Savilian Library at Oxford. It is entitled λῆμμα τῦ ἀναλυομένυ τόπυ; and an accurate copy of it, from the Parifian MSS. is placed in Dr. SIMSON's PAPPUS, from which a tranflation by the Doctor is added.

* Four only in COMMANDINE's tranflation are numbered as Propofitions, though it feems that five Lemmata were intended. And it is alfo to be remarked, that the four laft lines of fol. 300. av COMMAND. though printed in the character of the Commentary, are part of the text, of which the Greek is to be found in Savil. MS. No. 3. Dr. SIMSON, however, does not appear to have confidered thefe Lemmata; at leaft he has left no notes refpecting them. See the Memoir, p. 48.

" Propofitio PAPPI ALEXANDRINI quam ex COD. MSS.†

" Bibliothecæ Parifienfis paucis abhinc Septimanis exfcripfit

" D^{nm} JACOBUS MOOR, collega meus, 7mo Nov^{ris} 1748.

" Si latera circa angulum rectum trianguli BAC, (Fig. 10.)

" viz. BA AC, fecentur in K, L, ita ut tam BK ad KA quam AL

" ad LC fit ut BA ad AC, et jungantur BL CK, fibi mutuo

" occurrentes in F ; et ad bafim BC ducatur AFG, erit AG ad

" BC perpendicularis.

" Ducatur enim CE parallela ipfi BA occurratque ipfis BF,

" AG, (productis) in E, H ; et puta verum effe theorema, fc.

" angulum AGB rectum effe : æquiangula igitur funt BAC

" ACH triangula, quia ABC angulus æqualis eft ipfi GAC feu

" HAC (8 6.) et recti funt anguli BAC, ACH; ut igitur BA ad

" AC, ita AC ad CH. Eft autem propter parallelas EC ad CH

" ut (BK ad KA hoc eft ex hypothefi, ut BA ad AC hoc eft ut)

" AC ad CH: æquales igitur funt EC, CA. Et propter parallelas

" eft AL ad LC ut BA ad CE, hoc eft ad AC, quod quidem

" verum eft ex hypothefi.

" Componetur vero ita. Eadem manente conftructione,

" quoniam ex hypothefi eft BA ad AC ut AL ad LC, hoc eft

" ut BA ad CE; erunt AC, CE, æquales. Eft itidem ex hypo-

" thefi BA ad AC ut (BK ad KA hoc eft ut) CE feu AC ad

" CH : ergo (6. 6.) æquiangula funt triangula BAC, ACH, et

" angulus ABC æqualis ipfi HAC feu GAC, et in triangulis ABC,

" GAC communis eft angulus ACB, ergo reliquus AGC æqualis

" eft reliquo BAC, fc. angulo recto. Q. E. D.

" N. B. In Prop. PAPPI facile oftenditur AK, AL, equales effe."

† " Viz. No. 2368, et No. 2440.

In the fame MS. volume from which the preceding Propofition is copied, there is the following different ftatement of it :
[The fame figure.]

" Si fit triangulum ABC rectum habens angulum BAC,
" ducatur autem BD parallela ipfi AC æqualis autem rectæ
" BA; et CE parallela ipfi BA æqualis autem ipfi AC ; et CD
" BE jungantur fibi mutuo occurrentes in F, et juncta AF
" occurrat bafi BC in G, erit AG perpendicularis ad BC."

This Propofition, which is equivalent to the preceding, was
fuggefted by the figure of the 47 Prop. 1. EUCL. and had been
demonftrated by Dr. SIMSON, before he met with the other.
There is no intimation in the manufcripts hitherto examined
of the purpofe for which this Propofition was placed at the end
of this book ; and though it is called a lemma of the τόπος
ἀναλυομένος, there is no apparent connection between it and
any of the treatifes defcribed in this book, or with any of the
Lemmata belonging to them ; but from its being found in fo
many manufcripts of PAPPUS, there is a prefumption of its
having been placed in the *Mathematical Collections* by him.

§. VII. *The Eighth Book.*

THE object of this laft book* is to give fome account of the
ancient mechanics; and though a curious document of the
ftate of that branch of fcience in the time of PAPPUS, yet from

---

* The title from MS. BULL. is " Πάππου 'Αλεξανδρίνου συναγωγῆς ἦον περιέχει δὲ
" μηχανικὰ προβλήματα σύμμικτα§ ἕτθερά."

§ σύμμικτα is wanting in MS. Savil. No. 3.

·the great improvement both in the theory and practice of mechanics in modern times, it is comparatively of little value.

.The original genius of ARCHIMEDES was diftinguifhed in this department, as appears in part even from the portions of his numerous writings which have been preferved, and alfo from the many references to his mechanical inventions in this book of PAPPUS, and in many other ancient writers. A confiderable part of this book is employed in defcribing what were then, and ftill are, called the Mechanic Powers, and the moft obvious combinations of them for the common purpofes of life; efpe-cially for raifing up and for drawing very great weights.* PAPPUS avows its being chiefly borrowed from HERO, a diftin-guifhed geometer and mechanician, whofe works he frequently quotes, and fome of which ftill remain.† There is a long preface, which from fome ftatements of the mechanical notions, and of the arts of that period, becomes curious. It contains alfo fome obfervations on the utility of mechanics, and of the connection of mechanics with geometry; in it alfo are diftinguifhed feveral branches of that fcience, with notices of treatifes on it which are loft. Befides many references to HERO, there are very ample teftimonies in this preface of the fame of ARCHIMEDES, from his mechanical writings

* Towards the end of the book are defcriptions of the ancient machines for thefe purpofes; and in the MS. copies are defigns of fuch machines made from thefe defcriptions. It may be obferved, however, that thofe drawings are different from each other, and from the engravings in COMMANDINE, which no doubt had been copied from fketches in the MS. which he followed.

† This is the elder HERO, of Alexandria, who is fuppofed to have lived about fifty years after ARCHIMEDES.——GER. VOSSIUS.

and inventions,‡ in many of which his geometrical and
arithmetical fcience was employed. But as has already been
remarked, the chief value of this book is from the information which it affords of the ftate of mechanical fcience in
the age of PAPPUS. I muſt obferve alfo, that though Dr.
SIMSON makes a few corrections on this book, he feems never
to have confidered it particularly, and therefore it is unneceffary
to give a more particular account of its contents.§

§. VIII. *Conclufion.*

FROM the fhort account which has been given of the feveral
books of PAPPUS, the mifcellaneous nature of the Collection
is fufficiently apparent. Several ancient mathematicians are
mentioned by him, of whom no other notice remains; and

‡ In Prop. 10th, he mentions the well-known obfervation of ARCHIMEDES, that
with a fixed ftation he could move the earth. And the refolution of the general
mechanical problem, " to move a given weight with a given power," he calls the 40th
invention of ARCHIMEDES.

§ It is proper, however, to mention an obfervation of Dr. SIMSON on Prop. 14 of
this book : " Addatur hic (poft verbum *quidem* in linea quarta hujus paginæ, vir. fol.
320. b.) " *Geometrice* ut dudum emendavit GREGORIUS a ST°. VICENTIO, pag. 291."
Quad. Circuli.——In the fame place is alfo the following note, referring to the *Codex*
BULL.; he fays, " In quo poft verba ὀργανικὸς εὑρεῖν, legitur μεθοδεύεται δὲ τὸ τ΄ό·ὧεν τὰ·κι.
" Verbo vero μεθοδεύεται fignificatur conftructio geometrica." GREGORY alfo
makes another remark on COMMANDINE's Note B. in the fame page of PAPPUS, (1588.)
The laft fentence of the enunciation of Prop. 12, (fol. 317. a) in COMMANDINE's
tranflation is " Invenitur autem methodo inveftigata hoc pacto ;" but it ought to
have been " inveftigabitur autem geometrice ita ;" the Greek (MS. BULL.) being
" ἰαρονᾶται μεθόδιον ὅτως." Dr. SIMSON alfo remarks, refpecting the laft fentence of
COMMANDINE's Note B. on Prop: 14, viz. ' Mirum eft PAPPUM,' &c. " Non mirum
" quoniam ex elementis conicis fatis hujus demonftratio patet."

z 2

many curious particulars are detailed refpecting the ancient
ftate of mathematics, and of the problems and theorems which
engaged the attention of the geometers of thofe times. It is
written with only a general attention to method; there is little
uniformity in the ftile, and the work has probably been compofed at different times. Some propofitions are demonftrated
minutely, and rather diffufely, while in others many important
fteps are omitted,* as is often remarked by COMMANDINE, and
thefe omiffions he has generally fupplied. As a further proof of
this character of the *Collections*, the very frequent repetitions in
different parts of it, of the fame Propofitions, fometimes with
the fame, and fometimes with different demonftrations, may be
mentioned.§ It is, however, a moft interefting work, and more
efpecially from the account of the analytical geometry of the.
ancients in the feventh book; an edition of the Greek, from
a collation of the many manufcripts which are known ftill to
exift in the libraries of Europe, would be moft acceptable to
all the admirers of the elegance and accuracy of ancient
geometry; and for fuch an undertaking Dr. SIMSON's notes

---

* Dr. SIMSON makes a remark to this purpofe, *Opera Reliqua*, p. 526, at the end
of Prop. 76. Many of the imperfections of PAPPUS, as they appear in the exifting
MSS., muft be attributed to the tranfcribers.

§ The following examples may be remarked. The two ftatements of the ancient
claffification of problems, in b. iii. fol. 4. b. and b. iv. fol. 60. The two accounts of
the Conchoid in thefe two books. The two folutions of the Deliacal Problem, b. iii.
fol. 7, and b. viii. Prop. 2, nearly in the fame words. In Dr. SIMSON's edition of the
Sectio Determinata, feveral duplicates are mentioned. Other repetitions alfo might
be pointed out, fuch as Props. 31, 57, 178, 194, book vii. Alfo Props. 87, 153, 206,
in the fame book. Alfo Props. 23, 58, 193, b. vii. Several of thefe are remarked by
Dr. SIMSON, in his notes, both printed and unpublifhed. See *Opera Reliqua*, pp.
409, 410.

would be highly valuable.‡ He does not seem to have taken a regular and accurate survey of the whole, with any view to the immediate completion of such a work ; nor has he considered every passage which would require correction or explanation ; yet his emendations and illustrations, several of which have been mentioned in this Memoir, will be an important repository of materials for the assistance of any future editor.

It is proper likewise to observe, that among Dr. SIMSON's notes are several generalizations of Propositions in PAPPUS, and also some connected Propositions, which, though valuable in themselves, may probably not be all considered as properly belonging to a new edition of that author. Several also of the Doctor's notes contain corrections of COMMANDINE's commentary, most of which might be omitted in conducting a new edition ; of which the proper object must be, to give the text of PAPPUS free from errors, and to introduce such comments only as are necessary for explaining real difficulties, and for filling up material deficiencies in the demonstrations.

It may be also observed, that from what is known respecting several of the remaining MSS. and the similarity of their defects, it is much to be apprehended that few things of material importance, beyond what Dr. SIMSON has already remarked, are likely to be discovered; though an accurate examination of these MSS. would certainly be most desirable.

‡ By comparing Dr. SIMSON's MS. notes on the Lemmata of the seventh book with his corrected edition of these Lemmata in his restorations of the *Loci Plani Sectio Determinata*, and *Porisms*, his taste and judgment may be remarked in selecting only those notes which were necessary, and omitting the others which he had written on his first review of these Propositions.

After every aid is obtained from remaining MSS. the ability and intelligence of an editor must be depended on for the best use of exifting materials; and it is to be confidered, that, in geometry, the nature of the fubject, and the train of argument, may often enable an editor to fupply the defects and correct the errors of manufcripts, to which, indeed, thofe on mathematical fubjects appear to be more liable than any others.

The copy of COMMANDINE's PAPPUS which Dr. SIMSON ufed, and in which he wrote his notes, was the firft edition PISAURI, 1588, and VENETIIS, 1589, and to this all quotations in this memoir refer.

It is neceffary alfo to mention, that a few years after the Doctor's death, an application was made to his executor Mr. CLOW, on the part of the Delegates of the Clarendon Prefs, in the Univerfity of Oxford, for a tranfcript of all the MS. notes in this volume, as they were left by the Doctor. This was readily complied with, on the reafonable condition, that whatever notes and corrections of Dr SIMSON's might be adopted in a new edition of PAPPUS, they fhould be particularly diftinguifhed and acknowledged as his. Some time after, Mr. CLOW depofited this valuable *legacy* from his deceafed friend, in the library of the College of Glafgow ; and having had, by the favour of the Principal and Profeffors, every convenience for confulting it, I have been enabled to give from it fome interefting particulars of Dr. SIMSON's geometrical ftudies.

§. IX. ˙Of the *MSS.* of Pappus.˙

A Confiderable number of manufcripts of the collections of Pappus are to be found in various libraries of Europe, but all of them which have been examined, are mutilated ; and contain many errors, from the ignorance or carelefsnefs of the tranfcribers.†

Commandine appears to have had the ufe of only one manufcript, of which the hiftory and fate are unknown; but from his numerous corrections of it in his comm ntary, in which he generally inferts the erroneous paffage of the MS. a judgement may be formed of its very deficient ftate. In it, as in feveral other MSS.‡ the two firft books are entirely wanting.   ·

•In feveral MSS. however, a portion of the fecond book (about one half) is preferved. This was found in one of the Savilian MSS. viz. No. 9, and publifhed by Dr. Wallis, in 1688, with learned notes. It is underftood alfo, that one

---

† Dr. Simson, in a note on a Propofition, of the feventh book of Pappus, remarks, "; non; panca autem in hac Propofitione, in eo codice (Parifienfe Regio fc.) " vitiata funt, ut in omnibus fere Parfi Propofitionibus, et, ut videtur, in omnibus " manufcriptis."

‡ The fame is to be remarked of the Savil. MS. No. 3 ; alfo of a MS. of Pappus in the Neapolitian Royal Library, mentioned by Fabricius *Bibl. Græc.* tom. v. page 790, Hamburgb, 1798, &c. In the MS. Bullialdi, afterwards to be mentioned, the two firft books alfo are wanting. Ger. Vossius, *de Scientiis Math.* cap. xvi. 7, when ftating the books of the *Mathematical Collections* adds, " fed duo primi videntur. " deperiffe."

of the Parifian MSS. No. 2368, has this portion of the fecond book, beginning at the fame words as the fragments in the Savilian MS.‡ From a MS. of Pappus, which belonged to Joseph Scaliger, the portions of the preface to the viith book, defcribing the *Sectio Rationis* and *Sectio Spatii*, were publifhed in Greek by Snellius, in his reftoration of thefe books, in 1607 ; and his edition of the Greek preface to the *Scctio Determinata*, 1608, was probably from the fame MS.

The MS. of Pappus which was in the library of Isaac Vossius, and was carried from England to the Univerfity of Leyden, had alfo about one half of the fecond book*; and probably, the precife portion which is in the other MSS. now mentioned.

In the edition of the *Bibliotheca Græca*, publifhed by Harles, vol. ix. p. 170, is a catalogue of the MSS. of Pappus known to that writer, including thofe now mentioned and feveral others. I fhall in this place therefore mention only two more, not taken notice of by him. In the year 1795, J. Gugl. Camerer publifhed an edition of the Apollonius Gallus by Vieta ; and prefixes to it the preface and Lemmata belonging to the Tactions, in Greek. For this purpofe he made ufe of three MSS. the two in the Parifian library No. 1440, and No. 2368; and another in the Strafburgh library, not mentioned in Fabricius; but from the very fhort references

‡ A MS. copy of this fragment of the fecond book, (with the third book also,) taken from this Parisian MS. is in the Savilian Library at Oxford; and from a note at the end of this copy, it appears that the Parisian MS. was written in 1562.

* See the account of Isaac Vossius in the *Biog. Brit.* This MS. is mentioned in the Oxford Catalogue of Manufcripts, No. 2126.

to that MS. nothing can be pronounced refpecting its hiftory and value.

There is alfo at Edinburgh an elegant manufcript of five books of PAPPUS, viz. the third, fourth, fifth, fixth, and eighth; but unfortunately the feventh, the moft valuable portion of the work, is wanting. This MS. was purchafed at Paris, in 1748, by Dr. JAMES MOOR, then Greek profeffor at Glafgow, from Mr. DE MAIRAN; of the Academy of Sciences; and on the firft page of it is written *D'Ortous de Mairan*, probably by his own hand. DE MAIRAN had informed Dr. MOOR that it had belonged to BULLIALDUS; and on that account Dr. SIMSON, who had it for fome time in his poffeffion, and took many notes from it, calls it *Codex* BULLIALDI.† Like the other MSS. of PAPPUS it abounds with errors; but Doctor SIMSON obtained from it feveral improved readings and corrections, which he has remarked in his copy of COMMANDINE's tranflation. Some

† Dr. SIMSON, in his copy of COMMANDINE's PAPPUS, has the following notices of this MS. which, without doubt, he had from Dr. Moor; and which may therefore be confidered as authentic. In fol. 320. b. (PAPPI, edit. Pif. 1588.) " 3tie Nov͟ili A. D.
" 1748. Predictam GREGORII (a Sanct. Vinc.) emendationem veram effe, oftendit
" codex manufcriptus elegantiffimus, quem tribus abhinc diebus Parifiis huc attulit
" D. JACOBUS MOOR, collega meus doctiffimus, &c." .
Alfo in fol. 1. b. PAPPI, " Hifce litteris adfcriptis emendavimus tum fchema
" COMMANDINI tum MSti GRÆCI quem Dom. *D'Ortous de Mairan* dicit fuiffe
" BULLIALDI. R. S."
In a detached paper, referring to fol. 77. b. PAPPI, is a note in Dr. SIMSON's hand,
" 25 Martii 1750, inveni hæc ita effe in codice GRÆCO qui BULLIALDI fuerat,
" quemque collega meus D. JACOBUS MOOR, a domino Mairan emptum Parifiis huc
" (Glafguam) attulit."

A A

of them are curious and important, and all merit the attention
of any future editor of that work.

Some peculiarities in this MS. may be mentioned, for the
fatisfaction of thofe who may be inclined to confult it.

At the end of the third book are five pages of MS. not
in COMMANDINE's edition, which contain fome variations of
the folution of the Deliacal problem, in addition to thofe
contained in the beginning of that book.§

At the end of the fourth book is a very imperfect fketch
of a Propofition, not in COMMANDINE, which has already
been mentioned in the preceding account of this fourth book.

In this MS. are many corrections on the margin, and
fome even in the text, by a later hand; and among them
are a great number of the emendations propofed by COM-
MANDINE of the very fame errors, in the MS. ufed by him,
from which a connection between thefe two MSS. may be
inferred. Many of thefe emendations are fervilely copied,
retaining even the miftakes into which COMMANDINE, in a
few of them, had fallen.

This MS. though elegantly written, has been copied by
a perfon totally ignorant of the fubject; of which the number
of grofs errors in its firft ftate is a fufficient proof. It may be
remarked, alfo, that at the beginning of a Propofition, or of a
paragraph, there is ufually a red letter, and a new line; but

§ To thefe additional pages the following title is prefixed, " ἀλλως.—το δικαιον
" θεωρημα ἐν τῳ τριτω τῆς τῆ Παππου συναγωγης και τω ἀποδειξιν περιεχον, και τω ὀργανικην
" κατασκευην τῳ τι διπλασιασμῳ τῷ κυβυ και των δύο μεσων ἀναλογον."—It may be obferved
that the SAVIL. MS. No. 3, has a fimilar addition at the end of the 3d book.

frequently alfo this diftinction is made in the middle of a
fentence, while the beginning of Propofitions and fubjects
in other places is not diftinguifhed in that or in any other
manner. The enumeration of the Propofitions is often irregular, and many of the diagrams, though neatly drawn, are
altogether erroneous. The MS. was fold at Dr. Moor's death,
and was afterwards purchafed for the library of the Faculty of
Advocates at Edinburgh, in which the writer of this Memoir
had every facility for confulting that MS. and other books
in that great collection. From the well-known liberality
of the Curators of that eftablifhment, the aid of this MS.
will, without doubt, be moft readily given, when it can be
ufeful for preparing an edition of that interefting work in the
original language.

This MS. not having the feventh book, Dr. Moor procured
a copy of it to be taken from the MS. No. 2368, in the
Parifian library. This alfo for fome time was in Dr. Simson's
poffeffion, and he adopts from it fome corrections of Commandine's tranflation in his reftoration of the *Loci plani*, and *Sectio
determinata*; but, as was already mentioned, there is no reference to it in his expofition of the Porifms‡. I add with
regret, that this copy, in the difperfion of Dr. Moor's library,
feems to have been loft; and thefe particulars are ftated, as
they may perhaps facilitate the recovery of it from fome
obfcure fituation into which it may have accidentally fallen.

‡ In a note on Commandine's Pappus, the Doctor gives the following account
of this tranfcript, which he muft have got from Dr. Moor " Hunc autem librum
7mum Pappi ex eo codice (fcil. No. 2368, Reg. Bibl. Par.) defcripfit Dom Capiro-
nier, linguæ Græcæ in Academia Parifienfi Profeffor; fumptibus D. Jac. Moor,
collegæ mei doctiffimi. Schemata vero ejus depinxit D. Joa. Brisbane, M.D."

The two Savilian MSS. No. 3, and No. 9, have been repeatedly referred to. There is alfo in that library a tranfcript from
the Parifian MS. No. 2368, of a portion of the fecond book of
PAPPUS, (the fame as the fragment in MS. SAVIL. No. 9,
publifhed by Dr. WALLIS,) and the whole of the third book.
This tranfcript is not mentioned by Dr. WALLIS; and in the
following Appendix II. it is quoted as MS. SAVIL. B. From a
note at the end, it appears that the Parifian MS. No. 2368, was
written in 1562, by the direction of P. RAMUS.

APPENDIX II.

TWO paſſages of PAPPUS, on the ancient diviſion of geometrical lines into claſſes, have, in the preceding Memoir, been repeatedly alluded to. For the ſatisfaction of thoſe who may wiſh to conſider them accurately, the Greek, from the SAVILIAN MSS. and the MS. BULL. is printed in this Appendix, along with COMMANDINE's tranſlation. Some of the various readings in theſe MSS. are remarked, but without taking notice of either the ſmaller differences, which do not affect the ſenſe, or of ſome groſs errors, manifeſtly ariſing from the careleſſneſs of the tranſcribers.

Some of COMMANDINE's notes are alſo mentioned; but as Dr. SIMSON has not left any obſervations on theſe paſſages, it is not my purpoſe to enter into any examination or explanation of them, except only by pointing out the references to ſome Propoſitions, not explicitly ſtated either in PAPPUS, or in the commentary of his tranſlator.

PAPPI, MS. Savil. No. 3. Fol. 10, b.

"Ἃ μὲν ἴδει με προειπεῖν ἐστι ταῦτα. Παρ' οἷς
"δὲ κρίνειν σοί τε καὶ ἄλλοις τοῖς ἐν γεωμετρίᾳ
"γεγυμνασμένοις τὰ ὑπ' ἐκείνου προσγραφέντα∥
"περὶ τῆς κατασκευῆς, καὶ τὰ ὑφ' ἡμῶν ἐπε-
"νεχθέντα καλῶς ἔχειν ἡγοῦμαι, καὶ τὰ δόξαντα
"τοῖς ἀρχαίοις περὶ τοῦ προειρημένου προβλή-
"ματος ἐκθέσθαι. Καὶ πρῶτον εἰπὼν ὀλίγα
"περὶ τῶν ἐν γεωμετρίᾳ π‚οβλημάτων ἀρχὴν
"λαβὼν ἐντεῦθεν. Τῶν ἐν γεωμετρίᾳ προ-
"βλημάτων οἱ παλαιοί τρία γένη φασὶν εἶναι.
"Καὶ τὰ μὲν αὐτῶν ἐπίπεδα καλεῖσθαι, τὰ
"δὲ στερεά, τὰ δὲ γραμμικά. Τὰ μὲν οὖν δι'
"εὐθείας καὶ κύκλου περιφερείας δυνάμενα
"λύεσθαι λέγοιτο ἂν εἰκότως ἐπίπεδα. Καὶ
"γὰρ αἱ γραμμαὶ δι' ὧν λύεται τὰ τοιαῦτα
"προβλήματα, τὴν γένεσιν ἔχουσιν ἐν ἐπιπέδῳ.
"ὅσα δὲ προβλήματα λύεται παραλαμβανο-
"μένης εἰς τὴν γένεσιν μιᾶς τῶν τοῦ κώνου
"τομῶν, ἢ πλειόνων, ταῦτα στερεὰ κέκληται.
"πρὸς γὰρ τὴν κατασκευὴν ἀναγκαῖόν ἐστι
"χρήσασθαι στερεῶν σωμάτων† ἐπιφανείαις.
"Λέγω δὲ ταῖς κωνικαῖς. Τρίτον δ' ἔτι κα-
"ταλείπεται γένος, ὃ καλεῖται γραμμικόν.

" Quæ igitur me præmisisse opor-
" tebat, hæc sunt. Itaque omittens
" explicare et tibi, et iis, qui in geo-
" metria exercitati sunt, ea, quæ ille
" scripsit de constructione, et quæ
" nos objecimus; optimum fore ju-
" dicavi, si exponerem quid antiqui
" de dicto problemate senserint: et
" primum nonnulla dicerem de pro-
" blematibus, quæ in geometria con-
" siderantur, inde sumpto initio.
" Problematum geometricorum an-
" tiqui tria genera esse statuerunt, et
" eorum alia quidem *plana* appellari,
" alia *solida*, alia *linearia*. Quæ igitur
" per rectas lineas, et circuli circumfe-
" rentiam solvi possunt, merito *plana*
" dicantur; etenim *lineæ*, per quas
" ejusmodi problemata solvuntur, in
" plano ortum habent. Problemata
" vero quæcunque solvuntur, assumpta
" in constructionem aliqua coni secti-
" one, vel pluribus, *solida* appellantur;
" namque ad constructionem necesse
" est solidarum figurarum superfici-
" ebus, nimirum conicis, uti. Restat
" tertium genus, quod *lineare* appel-
" latur. Lineæ enim aliæ præter jam

" γραμμαὶ γὰρ ἕτεραι παρὰ τὰς εἰρημένας εἰς
" τὴν κατασκευὴν λαμβάνονται ποικιλωτέραν
" καὶ μεταπλασμίνην§ ἔχουσαι τὴν γένεσιν.
" ὁποῖαι τυγχά,* (ουσι ἀι ἴλικις) καὶ τετρα-
" γωνίζουσαι καὶ κογχλοειδῆς,‡ καὶ κισσοει-
" δῆς, πολλὰ καὶ παράδοξα περὶ αὐτὰς ἔχουσαι
" συμπτώματα. Τοιαύτης δὴ τῆς διαφορᾶς
" τῶν προβλημάτων οὔσης, οἱ παλαιοὶ γεωμέτραι
" τὸ προειρημένον ἐπὶ τῶν δύο εὐθειῶν πρόβλημα
" τῇ φύσει στερεὸν ὑπάρχον, οὐχ οἷοί τε ἦσαν
" κατασκευάζειν τῷ γεωμετρικῷ λόγῳ κατα-
" κολουθοῦντες. ἐπεὶ μηδὲ τὰς τοῦ κώνου τομὰς
" ῥᾳδιο τὸν ἐπιπέδῳ γράφειν ἦν." [ὡς δὴ δύο
" δοθεισῶν εὐθειῶν ἀνίσων, δύο μέσας ἀνάλογον
" λαβεῖν ἐν συνεχῇ ἀναλογία] Τοῖς δὲ ὀργάνοις
" μεταλαβόντες αὐτὸ θαυμασίως εἰς χειρουργίαν
" καὶ κατασκευὴν ἐπιτήδειον ἤγαγον. ὥς ἐστιν
" ἰδεῖν ἀπὸ τῶν φερομίνων αὐτοῖς συνταγμάτων.
" λέγω δ' ἐν τῷ ἐρατοσθένος μεσολάβῳ, καὶ
" τοῖς φίλωνος καὶ ἥρωνος μηχανικοῖς, ἢ κατα-
" παλτικοῖς. οὗτοι γὰρ ὁμολογοῦντες στερεὸν
" εἶναι τὸ πρόβλημα, τὴν κατασκευὴν αὐτοῦ
" μόνον ὀργανικῶς πεποίηνται, συμφώνως ἀπολ-
" λωνίῳ τῷ περγαίῳ. ὃς καὶ τὴν ἀνάλυσιν
" αὐτοῦ πεποίηνται διὰ τῶν τοῦ κώνου τομῶν.
" Καὶ ἄλλοι διὰ τῶν ἀρισταίου τόπων στερεῶν,

" dictas in conftructionem affumun-
" tur, varium, et tranfmutabilem or-
" tum habentes, quales funt helices,
" et [quas Græci τιτρα,ωνίζωσι appel-
" lant, nos] quadrantes [dicere poffu-
" mus,] conchoides, et ciffoides, qui-
" bus quidem multa, et admirabilia
" accidunt. Cum igitur tales fint pro-
" blematum differentiæ, antiqui geo-
" metræ problema ante dictum in
" duabus rectis lineis, quod natura
" folidum eft, geometrica ratione
" innixi conftruere non potuerunt;
" quoniam neque coni fectiones facile
" eft in plano defignare: inftrumentis
" autem ipfum in operationem manu-
" alem, et commodam, aptamque
" conftructionem mirabiliter tradux-
" erunt, quod videre licet in eorum
" voluminibus, quæ circumferuntur,
" ut in ERATOSTHENIS mefolabo,
" in PHILONIS, et HERONIS me-
" chanicis et catapulticis. Hi enim
" afferentes problema folidum effe,
" ipfius conftructionem inftrumentis
" tantum perficerunt, congruentur
" APOLLONIO Pergæo, qui et refolu-
" tionem ejus fecit per coni fectiones:
" alii per locos folidos ARISTÆI:
" nullus autem per ea, quæ proprie

§ βιβιασμίνη, idem.

* In MS. Savil. No. 3, quæ inter uncos, rubro liquore exaratæ a secunda manu.—In altero codice Savil. " ἀι ἴλικις" absunt. In MS. BULL. τυγχάνουσι ἀι . . . . καὶ.

‡ κογχοειδῆς, MS. B. SAVIL.

‖ Quæ inter uncos, in COMMANDINI verfione omittuntur; fed quamvis fupervacanea, forfan non erronea.

" οὐδεὶς δὲ διὰ τῶν ἰδίως ἐπιπέδων καλουμένων.
" Νικομήδης δὲ καὶ λόγῳ διὰ κογχλοειδοῦς*
" γραμμικῶς, δι᾽ ἧς καὶ τὴν γωνίας ἐτριχοτό-
" μησεν. ἐκθησόμεθα οὖν τίσσαρας αὐτοῦ κατα-
" σκευὰς, μετά τινος ἐμῆς ἐπεξεργασίας. ὧν
" πρώτην μὲν ἐρατοσθένειον, δευτέραν δὲ τὴν
" τῶν περὶ νικομήδη, τρίτην δὲ τὴν τῶν περὶ
" ἥρωνα μάλιστα πρὸς τὰς χειρουργίας ἁρμό-
" ζουσαν, τοῖς ἀρχιτεκτονεῖν βουλομένοις. Καὶ
" τελευταίαν τὴν ὑφ᾽ ἡμῶν ἀνευρημένην. στερεὸν†
" γὰρ παντὸς, ἕτερον στερεὸν ὅμοιον τῷ δοθέντι
" κατασκευάζεται πρὸς τὸν δοθέντα λόγον.
" ἐὰν δύο τῶν δοθεισῶν εὐθειῶν δύο κ.ε.ξ μέσαι
" κατὰ τὸ συνεχὲς ἀνάλογον ληφθῶσιν, ὡς
" ἥρων ἐν μηχανικοῖς καὶ καταπελτικοῖς.

" plana appellantur. At NICOME-
" DES, et ratione illud fecit per lineam
" conchoidem, per quam et angulum
" tripartito divifit. Exponemus igitur
" quatuor ejus conftructiones una cum
" quadam noftra tractatione. Qua-
" rum prima quidem eft ERATOS-
" THENIS, fecunda NICOMEDIS,
" tertia HERONIS, maxime ad ma-
" nuum operationem accommodata,
" iis qui architecti effe volunt. Ultima
" autem eft a nobis inventa : folido
" enim quocunque dato, alterum fo-
" lidum dato fimile conftruitur ad
" datam proportionem, fi duabus
" datis rectis lineis, duæ mediæ in
" continua analogia affumantur, ut
" inquit HERO in mechanicis ét
" catapulticis.

* κογχοειδὴς, MS. SAVIL. B.

† στερεῶ δεδομένω. MS. BULL. et ut videtur MS. quo utebatur COMMANDINUS.

‡ Deeft καὶ MS. B. SAVIL. et MS. BULL.

PAPPI, MS. SAVIL. No. 3. Fol. 86, b.

"Τὴν δοθεῖσαν γωνίαν εὐθύγραμμον εἰς τρία
"ἴσα τιμεῖν οἱ παλαιοὶ γεωμέτραι θελήσαντες,
"ἠπόρησαν, δι᾽ αἰτίαν τοιαύτην. τρία γένη φαμὲν
"εἶναι τῶν ἐν γεωμετρία προβλημάτων, καὶ τὰ
"μὲν αὐτῶν ἐπίπεδα καλεῖσθαι, τὰ δὲ στερεά,
"τὰ δὲ γραμμικά. τὰ μὲν οὖν δι᾽ εὐθείας καὶ
"κύκλου περιφερείας δυνάμενα λύεσθαι, λέγοιτ᾽
"ἂν εἰκότως ἐπίπεδα. καὶ γὰρ αἱ γραμμαὶ
"δι᾽ ὧν εὑρίσκεται τὰ τοιαῦτα προβλήματα
"τὴν γένεσιν ἔχουσιν ἐν ἐπιπέδῳ ὅσα δὲ
"λύεσθαι* προβλήματα, παραλαμβανομένης
"εἰς τὴν γένεσιν μιᾶς τῶν τοῦ κώνου τομῶν, ἢ
"καὶ πλειόνων, στερεὰ ταῦτα κέκληται. πρὸς
"γὰρ τὴν κατασκευὴν χρήσασθαι στερεῶν σχη-
"μάτων ἐπιφανείαις, λέγω δὲ ταῖς κωνικαῖς,
"ἀναγκαῖον. τρίτον δέ τι προβλημάτων† ὑπο-
"λείπεται γένος, τὸ καλούμενον γραμμικὸν,
"γραμμαὶ γὰρ ἕτεραι παρὰ‡ τὰς εἰρημένας εἰς
"τὴν κατασκευὴν λαμβάνονται, ποικιλωτέραν
"ἔχουσαι τὴν γένεσιν, καὶ βεβιασμένην μᾶλλον,
"ἐξ ἀτακτοτέρων ἐπιφανειῶν, καὶ κινήσεων ἐπι-
"πεπλεγμένων γενόμεναι, τοιαῦται δέ εἰσιν αἵ τε
"ἐν τοῖς πρὸς ἐπιφανείαις καλουμένοις τόποις
"εὑρισκόμεναι γραμμαὶ, ἕτεραι τε τούτων

"Antiqui geometræ datum angulum
"rectilineum tripartito fecare volentes
"ob hanc caufam hæfitarunt. Pro-
"blematum, quæ in geometria confi-
"rantur, tria effe genera dicimus; et
"eorum alia quidem *plana*, alia *solida*,
"alia vero *linearia* appellari. Quæ
"igitur per rectas lineas, et circuli
"circumferentiam folvi poffunt, me-
"rito dicuntur *plana*: *lineæ* enim per
"quas talia problemata inveniuntur,
"in plano ortum habent. Quæcun-
"que vero folvuntur, affumpta in
"conftructionem aliqua coni fectione,
"vel etiam pluribus, *solida* appellata
"funt, quoniam ad conftructionem
"folidarum figurarum fuperficiebus,
"videlicet conicis, uti neceffarium eft.
"Relinquitur tertium genus proble-
"matum, quod *lineare* appellatur;
"lineæ enim aliæ, præter jam dictas
"in conftructionem affumuntur, quæ
"varium et difficilem ortum habent,
"ex inordinatis fuperficiebus, et mo-
"tibus implicatis factæ. Ejufmodi
"vero funt etiam lineæ, quæ in locis
"ad fuperficiem dictis inveniuntur,

"ποικιλώτεραι, καὶ πολλαὶ τὸ πλῆθος ὑπὸ
" δημητρίου τοῦ ἀλεξανδρέως ἐν ταῖς γραμμικαῖς
" ἐπισἰάσεσι,καὶ φίλωνος τοῦ τυανέως, ἐξ ἐπιπ-
" λοκῆς πλικτοειδῶνξ τι καὶ ἑτέρων παντοίων
" ἐπιφανειῶν εὑρισκόμεναι, πολλὰ καὶ θαυμαστὰ
" συμπτώματα περὶ αὐτὰς ἔχουσαι. καὶ τινες
" αὐτῶν ὑπὸ τῶν νεωτέρων ἠξιώθησαν λόγου
" πλείονος. μία δέ τις ἐξ αὐτῶν ἐστιν, ἡ καὶ
" παράδοξος ὑπὸ τοῦ μενελάου κληθεῖσα γραμμή.
" τοῦ δὲ αὐτῶν γένους ἕτεραι ἕλικές εἰσι τετρα-
" γωνιζουσαί τι, καὶ κοχλοειδεῖς καὶ κισσοειδεῖς.
" δοκῆ δὲ πως ἁμάρτημα τὸ τοιοῦτον οὐ μικρὸν
" εἶναι τοῖς γεωμέτραις, ὅταν ἐπίπεδον πρόβλημα
" διὰ τῶν κωνικῶν ἢ τῶν γραμμικῶν ὑπό τινος
" εὑρίσκεται. καὶ τὸ σύνολον ὅταν ἐξ ἀνοικείου

" et aliæ quædam magis variæ, et
" multæ a DEMETRIO ALEXAN-
" DRINO [ἐν ταῖς γραμμικαῖς ἐπιςάσεσι,
" hoc eſt] in linearibus aggreſſioni-
" bus; et a PHILONE TYANEO ex
" implicatione [πληκτοειδῶν,*]et aliarum
" varii generis ſuperficierum inventæ,
" quæ multa et admirabilia ſympto-
" mata continent: et nonnullæ ipſa-
" rum a junioribus dignæ exiſtimatæ
" ſunt, de quibus longus ſermo habe-
" retur. Una autem aliqua ex ipſis
" eſt, quæ et admirabilis a MENELAO
" appellatur.

" Ex hoc genere ſunt lineæ helices,
" et quadrantes, et conchoides, et ciſ-
" ſoides: videtur autem quodam-
" modo peccatum non parvum eſſe
" apud geometras, cum problema
" planum per conica, vel linearia ab
" aliquo invenitur, et ut ſummatim

§ πληκτοειδῶν. MS. BULL. et ut videtur MS. COMMAND.

* Πλακτοειδῶν, or πληκτοειδῶν. This word, as appears from the application of it in Prop.
29. lib. iv. denotes a claſs of geometrical ſurfaces, produced by ſome preciſe rule, on which
mathematical reaſoning reſpecting ſuch ſurfaces may be founded. By COMMANDINE, and
alſo by VOSSIUS, they are ſuppoſed to have got this name from the complex motions by which
they are produced. The *Plecloeides*, in Prop. 29, (ſee fig. of that Prop. in PAPPUS) is
formed by the motion of the ſtraight line LKI paſſing through the ſtraight line BLN, to
which it is always perpendicular, and alſo through the line produced by the common ſection
of the conic ſuperficies, and of the ſuperficies called *Cylindroeides*, both mentioned in that
Propoſition. See note G. p. 104 and 105.

Among the ancients, geometrical curves and figures generally got names from ſome reſem-
blance which they were ſuppoſed to have to common objects. Such are the *Conchoid* and
Ciſſoid, and alſo the *Lunulæ*, or *Meniſcus*, the *Pelecoides*, the *Hippopeda*, and ſeveral others
mentioned by PROCLUS. (See PROCL. on Def. 4. 8. 1 Elem.)——It is poſſible that
πληκτοειδὴς may have had a ſimilar origin, not now to be traced; but the etymology implied in
COMMANDINE's tranſlation of it may be right.

"γίνεται λύεται,‖ διότι έστι τὸ ἐν τῷ ε̄ τῶν
"ἀπολλωνίου κωνικῶν, ἐπὶ τῆς παραβολῆς πρό-
"βλημα, καὶ ἐν τῷ περὶ τῆς ἕλικος ὑπὸ ἀρχι-
"μήδους λαμβανομένη στιρ ἀ νῦυσις ἐπὶ κύκλου,
"μηδενὶ γαρ προςχρώμενον στιρεῶ, δυνατὸν εὑρεῖν
"τὸ ὑπ' αὐτοῦ γραφόμενον ἐωρημα. λέγω δη
"τὸ τὴν περιφέρειαν τοῦ ἐν τῇ πρώτῃ περιφορᾷ
"κύκλου, ἴσην ἀποδεῖξαι τῇ πρὸς ὀρθὰς ἀγομένῃ
"εὐθεία τῇ ἐκ τῆς γενέσεως τῆς ἐφαπτομένης
"τῆς ἕλικος. τοιαύτης δὴ τῆς διαφορᾶς τῶν
"προβλημάτων ὑπαρχούσης, οἱ πρότεροι γεω-

"dicam, cum ex improprio folvitur
"genere, quale eft in quinto libro
"conicorum APOLLONII‡ problema
"in parabola: et in libro de lineis
"fpiralibus ARCHIMEDIS: affumpta
"folida inclinatio in circulo: fieri enim
"poteft, ut nullo utentes folido, pro-
"blema ab ipfo defcriptum invenia-
"mus.† Dico autem circumferentiam
"circuli in prima circulatione defcrip-
"tam demonftrare æqualem rectæ
"lineæ, quæ a principio lineæ fpiralis
"ad rectos angulos ducitur ei quæ eft
"circulationis principium, et a recta
"linea fpiralem contingente termi-
"natur. Itaque cum hujufmodi fit
"problematum differentia, antiqui

‖ λυηῖχι γινεσις, MS. SAVIL. No. 9.

‡ In COMMANDINE's time, the firft four books only of APOLLONIUS's *Conics* had been
difcovered; and he thence makes no obfervation on this criticifm of PAPPUS, refpecting a
Propofition in the fifth book. ALEXANDER ANDERSON, however, in his *Exercitationum
Mathematicarum Decas Prima*, (Paris 1619,) long alfo before the fifth, fixth, and feventh
books of APOLLONIUS were recovered from the Arabic, inferred from inveftigations founded
on the general defcription of that fifth book by PAPPUS, that the problem about the
Parabola, here alluded to by him, was truly a folid problem; and in *Exerc.* V. (page 24,)
he ftates it, and refolves it. In the fifth book of APOLLONIUS we now find the firft cafe of
Prop. 58. is equivalent to the problem of ANDERSON.

† The Propofition of ARCHIMEDES mentioned in this place is the 18th in his treatife
De Helicibus. In that Propofition the 7th is neceffary, and in that 7th Propofition (*De Helicibus*) the inclination alluded to is affumed.——COMMANDINE, in Note C. refpecting this
inclination, refers to Prop. 128. b. i. of VITELLO's *Optics.* That Propofition is indeed a plane
inclination, and is affumed in Prop. 5th of ARCHIMEDES's Treatife. But in Prop. 7. of the
fame treatife, which (as has been obferved) is ufed in Prop. 18th, a folid inclination is affumed,
the fame as Prop. 130. b. i. of VITELLO, in which an hyperbola is required. See alfo
ALHAZEN, b. v. Propp. 32, 33. In Prop. 42 of b. iv. of PAPPUS, is a *Locus ad Hyperbolam,*
which he ftates to have been affumed by ARCHIMEDES, in the folid inclination employed by
him in his treatife *De Helicibus.*

188 APPENDIX II.

"μετραι, τὸ προειρημένον ἐπὶ τῆς γωνίας προ-
"βλημα, τῇ φύσει στερεὸν ὑπάρχον διὰ τῶν
"ἐπιπέδων ζητοῦντες, οὐχ οἷοι ἦσαν εὑρίσκειν,
"οὐδέ πω‖ γὰρ* τοῦ κώνου τομαὶ συνήθεις ἦσαν
"αὐτοῖς. καὶ διὰ τοῦτο ὑπόρησαν. ὕστερον
"μέντοι διὰ τῶν κωνικῶν ἐτριχοτόμησαν τὴν
"γωνίαν εἰς τὴν εὕρεσιν χρησάμενοι τῇ ὑπο-
"γεγραμμένῃ νεύσει."

" geometræ problema jam dictum in
" angulo, quod natura folidum eft,
" per plana inquirentes invenire non
" potuerunt: nondum enim ipfis cog-
" nitæ erant coni fectiones, et ob eam
" caufam hæfitarunt. Poftea vero
" angulum tripartito diviferunt ex
" conicis, ad inventionem infra fcripta
" inclinatione utentes.†

‖ ὁ δὴ ται, MS. Savil. No. 9.

* γα‹ ἀι‹. MS. Bull, αι τε κινω τομαι συνιχιις ιισιν. MS. Savil. No. 9.

† Prop. 31, lib. iv. Pappi.

APPENDIX III.

A S the volume of Dr. Simson's posthumous works is not
in general circulation, it may be agreeable to the
Mathematical reader, to fee the Doctor's very improved
tranflation of the general defcription of Euclid's Porifms,
in the preface to the feventh book of Pappus, annexed to
the account of his labours. The mutilated detail of the
contents of the three books of Porifms, not being materially
altered by Dr. Simson, except in a few Propofitions reftored
by him, is not added.

" POST Tactiones in tribus libris habentur Porifmata Euclidis, collectio
" artificiofiffima multarum rerum quæ fpectant ad analyfin difficiliorum et
" generalium problematum, quorum quidem ingentem copiam præbet natura.
" Nihil vero additum eft iis quæ Euclides primum fcripferat, præterquam
" quod imperiti quidam qui nos præcefferunt feeundas defcriptiones paucis
" ipforum [fc. Porifmatum] appofuerunt. Cum vero unumquodque definitum
" habeat demonftrationum numerum,† ut oftendimus,‡ Euclides unam
" eamque maxime evidentem in fingulis pofuit. Habent autem fubtilem et
" naturalem contemplationem, neceffariamque et maxime univerfalem, atque
" iis quæ fingula perfpicere et inveftigare valent admodum jucundam   Specie

† " Forfan intelligit demonftrationes diverforum cafuum ejufdem Porifmatis ; vel propo-
" fitionum quæ funt ejus converfæ.
‡ " In quibufdam ex lemmatibus, ni fallor, ad Porifmata.

191   APPENDIX III.

" autem hæc omnia neque theoremata funt, neque problemata, fɛd mediæ
" quodammodo inter hæc naturæ, ita ut eorum Propofitiones poffunt vel ut
" theoremata, vel ut problemata formari. Unde factum eft, ut inter multos
" Geometras alii hæc genere theoremata exiftiment, alii vero problemata,
" refpicientes ad formam tantum propofitionis. Differentias autem horum
" trium melius intellexiffe veteres manifeftum eft ex definitionibus. Dixerunt
" enim theorema effe quo aliquid propofitum eft demonftrandum ; problema
" vero quo aliquid propofitum eft conftruendum ; Porifma vero effe quo aliquid
" propofitum eft inveftigandum. A Neotericis autem immutata eft hæc
" Porifmatis definitio, qui hæc omnia inveftigare haud potuerunt, fed
" Elementis hifce adhibitis, oftenderunt tantum quid fit quod quæritur, non
" autem illud inveftigaverunt. Et quamvis a definitione et ab ipfis rebus quæ
" traditæ funt redarguerentur, hoc tamen modo, ab accidente, definierunt.
" Porifma eft quod deficit hypothefi a Theoremate Locali [hoc eft, Porifma eft
" Theorema Locale deficiens five diminuta in hypothefi ejus.] Hujus autem
" generis Porifmatum Loca Geometrica funt fpecies, quorum magna eft copia
" in Libris de Analyfi ; ac feorfim a Porifmatibus collecta, fub propriis titulis
" traduntur, eo quod magis diffufa et copiofa fit hæc præ cæteris fpeciebus. E
" Locis enim quædam plana funt quædam folida, quædam linearia, et præter
" hæc funt Loca ad medietates [five a mediis proportionalibus orta.] Accidit
" hoc etiam Porifmatibus, Propofitiones habere concifas propter difficultatem
" multarum rerum quæ fubintelligi folent; unde evenit Geometras non paucos
" ex parte tantum rem perfpicere, dum ea quæ inter oftenfa magis neceffaria
" funt haud capiunt. Multa autem ex iis in una Propofitione minime com-
" prehendi poffunt, quia ipfe EUCLIDES non multa in unaquaque fpecie
" pofuerit, fed ut fpecimen daret multæ copiæ, pauca ad principium primi
" libri pofuit ejufdem omnino fpeciei cum uberrima illa [quæ in primo libro
" habetur] Locorum fpecie, ut decem fint numero. Quare has [Propofitiones
" fcilicet hujus fpeciei] una Propofitione comprehendi poffe animadvertentes,
" eam ita defcribimus.

' Si quadrilateri cujus anguli oppofiti vel ex adverfo, vel ad eafdem partes
' funt pofiti,* [lateribus productis] data fint in uno ipforum tria puncta

* " Primum horum PAPPUS vocat ἰσθιον, alterum παρυτλίον, de quibus vide Not. C.
" pp. 85, 86, 87.

'[interfectionum scilicet;] vel fi in quadrilatero cujus duo latera funt inter fe
' parallela [data fint duo puncta interfectionum in altera parallelarum;] caetera
' vero puncta praeter unum tangant rectam pofitione datam; etiam hoc tanget
' rectam Pofitione datam.'   " Hoc autem de quatuor tantum rectis dicitur
" quarum non plures quam duae per idem punctum tranfeunt. In quolibet
" vero propofito rectarum numero ignoratur, quamvis vera fit hujufmodi
" Propofitio, viz.

' Si quotcunque rectae occurrant inter fe, nec plures quam duae per idem
' punctum ; data vero fint puncta omnia in earum una, unumquodque autem
' punctum in alia tangat rectam, pofitione|| datam.' Vel generalius fic. ' Si
' quotcunque rectae occurrant inter fe, neque fint plures quam duae per idem
' punctum, omnia vero puncta [interfectionem scilicet] in earum unâ data fint;
' reliquorum numerus erit numerus triangularis, cujus latus exhibet numerum
' punctorum rectam pofitione datam tangentium ; quarum interfectionum fi
' nullae tres exiftant ad angulos trianguli fpatii [nullae quatuor ad angulos
' quadrilateri, nullae quinque ad angulos quinquelateri, &c. i. e. Univerfim, fi
' nullae harum interfectionum in orbem redeant] unaquaeque interfectio reliqua
' tanget rectam pofitione datam.'

" EUCLIDEM autem hoc nefciviffe haud verifimile eft, fed principia fola
" refpexiffe : nam per omnia Porifmata non nifi prima principia, et femina
" tantum multarum et magnarum rerum fparfiffe videtur. Haec autem juxta
" hypothefium differentias minime diftinguenda funt; fed fecundum diffe-
" rentias accidentium et quaefitorum.   Hypothefes quidem omnes inter fe
" differunt, cum fpecialiffi.nae fint : accidentium vero et quaefitorum unum-
" quodque, cum fit unum idemque multis diverfisque hypothefibus contingit.§

" Talia itaque inquirenda offeruntur in primi libri Propofitionibus ; (in
" principio feptimi habetur* diagramma huc fpectans) ' Si a duobus punctis
' datis inflectantur duae rectae ad rectam pofitione datam, abfcindat autem
' earum una à rectâ pofitione data fegmentum dato in ea puncto adjacens,
' auferet etiam altera ab aliâ rectâ fegmentum datum habens rationem.' i. e.
" Quod ad alterum segmentum habebit rationem eandem rationi quae ex
" hypothesi data est."   Deinde in fubfequentibus ; &c."

|| " Unum tangat unam, aliud tangat aliam rectam pofitione datam, et fic deinceps."
§ " Ex gr. Multa funt Porifmata quae diverfas hypothefes habent, fed quae omnia
" concludunt punctum aliquod tangere rectam pofitione datam ; vel rectam aliquam vergere
" ad punctum datum, &c."
          * " Non jam habetur."

ERRATA.

Page 18, line laſt of note, for 87 read 37

38, l. 6, note, for *rationum* read *rationem*

40, l. 5, note, for 280 read 270

89, l. 5, after μέτα place a period.

90, l. laſt of note, for γ read ȳ (3)

91, l. dele note, "*forſan* ν̀δομαϝοις"

94, l. 1 note, for *indeterminatum* r. *indeterminatus*

114, l. 22, for *letter after*, read *letter, after*

121, l. 4, for *into ; it* read *into it ;*

Page 131, l. 3, for *treatiſes all of which ;* read *treatiſes, all of which*

132, l. 18, for *aſcertained* read *determined*

136, l. 5, note, for *Geometry ;* read *Aſtronomy ;*

145, l. 1, note, for δ' ὁπἰϸ read δ', ὁπἰϸ

149, l. 12, for *plane*, read *plane*

170, l. laſt, for *ſame* read *fame*

175, l. laſt note, for *deperiſſe* read *deperiiſſe*

176, the firſt note, from page 180, unneceſſary.